Maximum Entropy Solutions to Scientific Problems

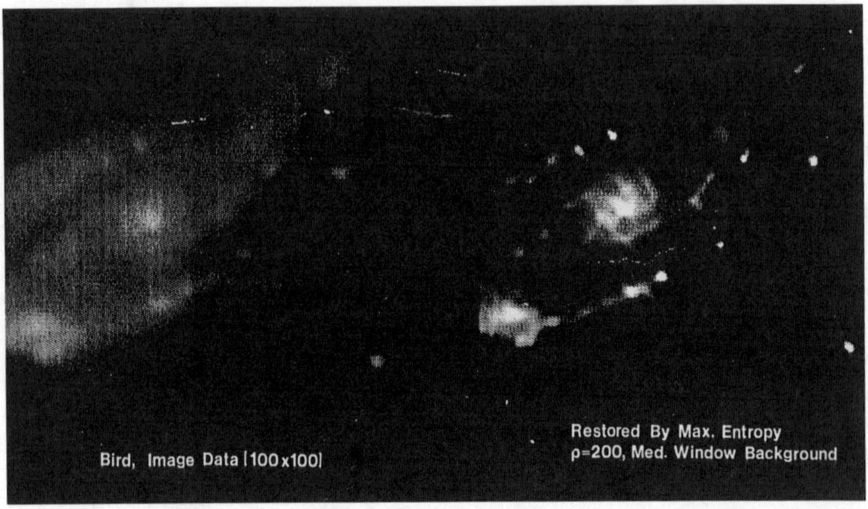

Maximum Entropy restoration of a fragment of the "Bird" galaxy consisting of resolvable stars, gaseous nebulae, and galactic arms as the foreground (object), superimposed on an independent, slowly varying background of faint stars and blurred by atmospheric turbulence. The image picture was taken by Kitt Peak National Observatory personnel. The ME procedure maximizes the entropy of the object subject to Poisson statistics for the image noise and the constraints of image intensity over 100 × 100 pixels. The background was estimated by sliding a median window over the image. Sharpness parameter ρ is defined as $(\Delta o / i_{max})(SNR)^2$ where Δo is the object intensity resolution limit, i_{max} is the maximum intensity level in the image and SNR is the signal-to-noise ratio at i_{max}. From "Restoring with Maximum Entropy. III. Poisson Sources and Backgrounds," by B. R. Frieden and D. C. Wells, *Jour. Opt. Soc. Am.*, 68, Jan. 1978, pp. 93–103. (Courtesy Dr. B. R. Frieden, Optical Sciences Center, University of Arizona, Tucson, AZ.)

Maximum Entropy Solutions to Scientific Problems

Robert M. Bevensee

P T R Prentice Hall
Englewood Cliffs, New Jersey 07632

Library of Congress Cataloging-in-Publication Data

Bevensee, R. M.
Maximum entropy solutions to scientific problems / Robert M. Bevensee.
p. cm.
Includes bibliographical references and index.
ISBN 0-13-563818-6
1. Entropy. I. Title.
QC318.E57B48 1993
507.6—dc20 92-3502
CIP

Editorial/production supervision: Mary P. Rottino
Cover design: Ben Santora
Prepress buyer: Mary E. McCartney
Manufacturing buyer: Susan Brunke
Acquisitions editor: Betty Sun

The publisher offers discounts on this book when ordered in bulk quantities. For more information, write:

Special Sales/Professional Marketing
Prentice-Hall, Inc.
Professional Technical Reference Division
Englewood Cliffs, New Jersey 07632

Printed in the United States of America

10 9 8 7 6 5 4 3 2 1

ISBN 0-13-563818-6

Prentice-Hall International (UK) Limited, *London*
Prentice-Hall of Australia Pty. Limited, *Sydney*
Prentice-Hall Canada Inc., *Toronto*
Prentice-Hall Hispanoamericana, S.A., *Mexico*
Prentice-Hall of India Private Limited, *New Delhi*
Prentice-Hall of Japan, Inc., *Tokyo*
Simon & Schuster Asia Pte. Ltd., *Singapore*
Editora Prentice-Hall do Brasil, Ltda., *Rio de Janeiro*

DEDICATION

In science we often seek the "most probable" solutions to problems when we lack information to solve them fully.

In art we value the rarer, "less probable," objects for our own selfish reasons.

In human relations we enjoy those unique persons who complement our personalities and who like—and love—us for our uniqueness.

This book is dedicated to three uniquely loveable people: my wife, Mae and my sons, Mark and Brendan.

Contents

Preface

Suppose you roll an unfamiliar pair of dice and bet someone on the outcome. What number would you bet on? Certainly not 2 or 12, either of which can come up only one way. You would intuitively bet on 7 because it can appear the most number of ways—six (6 and 1, 1 and 6, 5 and 2, etc.) Seven is the most probable number. In the language of this book it has the maximum entropy because the number of ways of realizing it is a maximum. Entropy is a measure of the number of ways of realizing an object. Even with some information on the outcome of previous rolls of the dice you could very reasonably bet on the most probable outcome, conditioned on that information, without any more abstract statistics.

This book solves various practical, underdetermined problems by several versions of the maximum entropy method. In an underdetermined problem one solves for the properties of an object from a distorted or incomplete image. The frontispiece shows a maximum entropy reconstruction of the Bird nebula which has been blurred by background noise. The number of data is insufficient to resolve the object exactly, therefore, we seek the most probable or maximum entropy object consistent with the data and with the fewest assumptions.

The full power of this method of reasoning is most impressive in the physical sciences, where a configuration of an astronomical number of molecules is obtained which can be realized the greatest number of distinct ways, consistent with the extremely sparse data of temperture, pressure, volume, and so forth. We are employing the idea that a "free" system tends to its most probable or maximum entropic state, the Second Law of Thermodynamics. Anyone who tries to keep a home or garden in order (low-entropic state) by fighting various disruptive influences that tend to randomize the environment (and thereby make it more probable and of higher entropy) is fighting the principle of the Second Law!

There is no one best way of solving underdetermined inverse problems to infer the details of objects from their incomplete or noisy images. The maximum entropy (ME) method seems to be one of the most, if not *the* most, powerful techniques for these problems. If you favor another method for a particular problem and I prefer the ME method, we can sometimes choose between them on the basis of known or "synthetic data" solutions of similar problems. However, if we can't decide that way you're certainly privileged to use your method. I can't argue very persuasively because neither of us knows the answer to the actual problem. But I may call you an intellectual Philistine and use my ME method—that's *my* privilege. Generally, it is obvious when the ME method is clearly inferior to another method of solution. Invariably, the ME method will predict, not additional information—for the method cannot generate information—but rather attributes of an object which corroborate or enhance the predictions of other analyses.

The ME method is flexible in that additional information can be fed to it and the most probable object enhanced. The new object may differ dramatically from the old, as illustrated by the frontispiece. However, we don't know the precise object anyway, so we shouldn't have any philosophically strong objections to choosing the updated, most probable object. If the ME object were to have attributes which we know, a priori, to be wrong, the data and knowledge conflict and must be reconciled.

It is surprising to find presumably unprejudiced observers quite unwilling to update their descriptions of objects with new data. Skilling and Gull made this statement in a recent book [1]: "It is ludicrous to suppose that our prejudices concerning the shape and structure of an astronomical object should depend on whether it is viewed in the optical or radio [frequency] bands"! To this we retort: if we observe by low-frequency radar, a flying "blob" and later discover by high-frequency radar that it is a missile, should we not assume it is a missile?

Inferring the characteristics of an object from incomplete data by the ME method is a bit like gambling that the ME object will be satisfactorily close to the true object. But this is not like gambling in a game of chance. Consider the savvy blackjack player. At the betting stage he or she considers (or should, if profit-motivated) *all possible* hands that he or she and the dealer might receive in the next game, and their probabilities. This gives him or her a net probability of win, and the person should bet accordingly. The ME blackjack player at some stage would try to deduce the *most probable* numerical pair of hands based on the number of distinct card combinations for realizing each numerical hand. He or she would bet high if the ME hand would beat the dealer's; otherwise, he or she would bet low. But the ME gambler wouldn't win as much because he or she would neglect, by the very nature of the ME method, the lower probability class of hands that would win more for the savvy player!

Many scientifically minded people cannot accept the ME viewpoint when it ignores a priori statistics about objects. Earth-modeling is an example. Suppose we want to construct a density model (object) of the earth from various data (the image) such as total mass and various moments of inertia. Obviously, there is an infinity of possible models, all consistent with the meager data, and with no statistics about a "large number" of earths. In such a situation, the usual ME viewpoint is that all models are equally probable. One then proceeds to derive, in effect, a density distribution that can be constructed the most ways with a large number of building blocks, consistent with the data. Even the ubiquitous normal (Gaussian) distribution need never be assumed, and it is often derived in the course of solving certain problems. We ME people are able, with the intellectual power which gave us the Second Law of Thermodynamics, to solve inverse problems that stump people who rely on classical statistics.

We illustrate the different viewpoints of a classical statistician (Mr. Classi) and a ME analyst (Mr. Maxi), who want to estimate the density throughout the earth. Both agree to assume, in first reasonable approximation, that the earth is spherically symmetric, which means the density depends on the distance, r, from the center. Call it $d(r)$, in grams per cubic centimeter. The mass and moment of inertia are known accurately from measurement. Classi speaks first.

Classi: (haughtily) To solve this "rather elementary" problem I will first set up a probability distribution for each d-value at each r, based on well-known physical laws.

Maxi: You mean the laws of gravitation and conservation of momentum. But how will you solve them? We agree that the earth spun off from the sun but we have only the vaguest idea of how the spatial distributions of mass and velocity evolved as the earth congealed. So, how can you predict the end result of a solid earth with a molten core?

Classi: Ah, but I have tremendous computer power! I'll assume a number N of possible initial distributions of mass and velocity and run my computer 1000 hours and look at the final N earth models. Then I'll set up a consistent probability distribution for $d(r)$.

Maxi: But why guess a probability distribution? Why not just index your various computer solutions by the letter i and estimate $d(r)$ at each r by the average value from all the N solutions.

Classi: Classical statistics would indeed allow this! So, I've solved the problem in a reasonable way, haven't I?

Maxi: (dryly) Aren't you forgetting the facts about the mass M and inertia J of the earth. Surely, if you average your solutions to get their average M and average J they're not going to agree with the data. Suppose you're off 10%. How are you going to correct for this?

Classi: (beginning to figit) I'll run my computer longer and solve more earth models until I get N of them with an average M and average J sufficiently close to the measured values.

Maxi: Admirable persistence, but is it practical? You say you need 1000 hours to get enough models to estimate $d(r)$ reasonably well. You'll probably need 100,000 hours to get enough models with accurate M and J. Isn't this getting rather expensive?

Classi: Expense is the prerequisite for accuracy. Have you a better way of solving the problem?

Maxi: Certainly. I like the idea of computing N earth models you say you can afford. To allow for the fact that the average density of the models is certainly in error at each r, and to enforce the correct mass and moment of inertia, do the following. Estimate the error in each $d(r)$ from your models as the standard deviation in your model $d(r)$'s. Then solve for the most probable $d(r)$-distribution with the constraints that it will be within the standard deviation at each r and also agree with the M and J values for the earth.

Classi: Your estimate of the error is certainly crude!

Maxi: You're welcome to improve it. The point is, you should solve for the most probable density distribution without assuming any statistics.

Classi: What! Give up my favorite probability distribution for this type of problem. That's asking a lot.

Maxi: How do you know which probability distribution nature chose when she congealed the earth?

Classi: (thoughtful) You say you want me to abandon any a priori probability distribution for $d(r)$ in favor of a "most probable" one. What do you mean by that?

Maxi: Simply, one that can be constructed the greatest number of distinct ways.

Classi: Sounds reasonable, but can you solve for such a distribution with the mass and inertia constraints without consuming a huge amount of computer time?

Maxi: Read Chap. 2 to find out two ways of doing it, depending on your preference.

Readers who want more elaboration on the intellectual controversy between the "Orthodoxians," who depend on classical statistics to solve problems, and the "Maxists," who create their own statistics as a by-product of their most-probable solutions, should read E. T. Jaynes's writings [2,3]. He, more than anyone else, deserves the title of "Dr. Maximum Entropy."

There is a related popular method of data inversion to reconstruct underdetermined objects, which goes by the name of Maximum Likelihood, abbreviated ML. I shall compare it philosophically to the ME method but will not solve any problems by ML. Many problems can be solved equally well by either method, and the choice is a subjective one. B. R. Frieden, professor of optical sciences, has discussed the essential features of the ML method in an intriguing article [4] in which he analyzes an experiment from both the ML- and ME-viewpoints. A special die with three possible outcomes—1,2, or 3—is thrown five (or 10) times and the average of these throws (the "image") turns out to be 2. The question is: what are the probabilities of each face turning up (the "object")? Assuming no other information, the ME answer is 1/3 for each number; that is, each number is equally likely to turn up each throw. The rather curious ML prediction is zero probability for each of the numbers 1 and 3 turning up and a probability of one for number 2 appearing. This means the number 2 is predicted to turn up *every throw*.

I cannot fault Frieden's ML analysis of this problem but I do dispute his implied claim that the ME result is wrong and the ML answer is right. There is no "right" and "wrong" answer to this problem; as the die is rolled more and more times the experimental probabilities of each number turning up approach the true probabilities. One cannot say very much after five or ten throws!

I believe it is best to start this book with a review of the ME derivation of the most common probability distribution in physics—the Maxwell-Boltzmann distribution, which gives us the "normal" or "bell-shaped" probability curve. I also derive the Fermi-Dirac distribution function for electrons and the Bose-Einstein distribution for photons, or light "particles." When I was a lowly graduate student at the University of Minnesota I enjoyed a course in thermodynamics given by A. O. C. Nier, who used the textbook *Thermodynamics,* by F. W. Sears [5]. My development follows closely Sears's lucid derivations. Chapter 1 continues with a brief description of various corrupting influences on the reconstruction of an object from incomplete data: statistical uncertainty in the true object relative to the ME one, and the various noises which can blur the ME reconstruction. Then I relate both the ME and ML inversion methods to the classical Bayes's Theorem of statistics, in which the role of prior object probabilities must be addressed. If the priors are known or estimated, the entropy expression can be modified to include them.

Chapter 1 then discusses two topics which involve both entropy and information. Here I distinguish between information (data, facts) and entropic information, which is the reduction in uncertainty of facts achieved by the message. I conclude from this that the entropy concept is the fundamental one, and equivalences between entropy and entropic information occur when information is defined in entropic terms! Then I look at a situation in which a group of investors interact with each other and try to increase their own money entropies. I conclude the net entropy of total capital may increase or decrease with time—because personal judgment is involved. Lastly, I point out an entropic reason for the popular belief that the rich grow richer while the poor grow poorer.

Chapter 2 jumps right into a practical problem in order to illustrate the ME method in operation. With virtually no information, we derive a density distribution for an assumed spherically symmetric earth which is consistent with the data and other estimates. Two varieties of the ME method are introduced. One is the Gibbsian method originated by the American physicist J. W. Gibbs at the turn of the century, in order to develop statistical mechanics and explain thermodynamics from it. A variation of this method was popularized in geophysical prospecting by J. P. Burg and his colleagues, the "Burgers," after Burg's classic paper in 1967 [6]. The other variant of the ME method is associated with the German physicist L. von Boltzmann of the last century, who introduced it into statistical mechanics in order to understand thermodynamics from a different philosophical viewpoint. Those who favor this ME method, the "Boltzmann," include this writer. However, the Gibbsian method is favored by Jaynes [2] and has advantages in certain problems which we shall not fail to observe.

Chapter 3 delves into the geotomographic class of problems in which we infer a parameter distribution of an object from a number of probing straight rays. These ME objects are formally equivalent to cross sections of human bodies examined with tomographic medical CAT scans, in which acoustic or other rays are used to probe in search of tumors. I specifically formulate the problem of resolving a two-dimensional slice of earth (for density, electrical conductivity, or perhaps dielectric constant) by tracing rays. I describe an algorithm which has proven generally successful for iterating to the Boltzmann ME parameter distribution, provided the number of data (rays) is less than about half the number of unknown parameters in the discretized cells in the slice. I show how the ME solution is a generalized inverse one, and why it tends to become numerically unstable when the number of data exceeds a certain criterion. Noise can be treated within this framework. The convergence to an ME object can often be improved by removing certain eigenvectors of small eigenvalues until the final ME solution is almost reached, and then reinserting the eigenvectors. Synthetic data examples illustrate the potential of this method.

In a practical problem, the probing rays bend or diffract through the spatially varying region (object) and the ME solution must allow for this. I indicate how the ME algorithm can be modified to update the raypaths each iteration as it converges to an ME solution.

The last two sections of Chap. 3 treat several very subtle and difficult topics in ME data processing, which concern the Confidence one has that a true parameter will be "close" to its ME value. This depends on the statistical distribution of possible objects or parameter values, subject to the constraints, about their ME values. Even without measurement noise there is what I call a *statistical uncertainty* of a parameter about its ME value.

The uncertainty in the entropy of an object can be described by Jaynes's Concentration Theorem; the uncertainty distribution of an object parameter can be quantized by Darwin-Fowler Theory. Of fundamental importance is the requirement that the object be quantized in units of a resolution level independent of any sources of noise. Measurements would determine this resolution level, and thereby, the statistical uncertainty of a parameter about its ME value. This uncertainty is quantified by the concept of *Assurance* that an object parameter will lie within a certain interval of its ME value with no noise present.

However, measurement noise usually swamps the statistical uncertainty. I quantify this situation and offer a formula for the variance of an object parameter about its ME value among the members of the object population which satisfy the constraints. This formula helps to estimate, without further computation, the confidence that the true value of an object parameter will lie within a specified interval of its ME value.

Chapter 4 outlines a successful Boltzmann tomographic inversion method for reconstructing a density distribution within a "slice" of an object from rapid recording of probing straight rays falling on a series of scan planes at different angles to the slice. We illustrate the method with a reconstruction of an electron beam density profile at the Lawrence Livermore National Laboratory. We show how the method can be modified to include prior knowledge and allow for missing data. An example shows how structural flaws in an object may be resolved more accurately. The same data inversion method could be employed to reconstruct human-body density cross sections from tomographic CAT scans.

Chapter 5 reviews ME spectral analysis from the Gibbsian viewpoint, using autocorrelation data. After a brief review of sampling theory we present R. N. McDonough's derivation [7] of the Burg ME power spectrum formula from the ME probability distribution of the time values of the function, and show the relationship to modeling by an autoregressive all-pole filter. Then, we review an alternative derivation of the Burg spectrum from a probability distribution of the power spectrum, due to E. Rietsch [8]. The time-frequency spectral analysis carries over by analogy to spatial array data processing in the space-wavenumber domains. The only difference is that the spatial autocorrelation function is Fourier-transformed into the frequency domain to give a power spectrum at the data-processing frequency. It is this power spectrum and the wavenumber spectrum as its Fourier transform which directly correspond to the autocorrelation function and its frequency spectrum, respectively, in time-frequency spectral analysis.

Then, we show the relationship of this ME autoregressive all-pole modeling to the "Prony Method" of pole-zero modeling. In Sec. 5.9, we examine the effect of white noise on the ME (super) resolution of two coherent, equiamplitude sinusoidal signals very close in frequency. S. L. Marple [9] has shown that one may superresolve the frequencies better than allowed by the classical Nyquist limit even for signal-to-noise ratios less than 1! His conclusions carry over by analogy to superresolution of noisy wavenumber spectra. The corresponding resolution of incoherent (random phase) sources is difficult but can be improved by removing the noise autocorrelation function from the spatial-array data processing.

In Chap. 6 I discuss coherent spectral analysis from the less popular Boltzmann point of view. For time-frequency spectral analysis, the data are the temporal values themselves, not their second-order autocorrelations. The constraint relations are the Fourier transforms between the data and their spectra (amplitude, not power, spectra now), and we write a power entropy involving the squared magnitude of these discrete spectra. This entropy does not have a global maximum at an extremal point but saddle point properties instead, relative to changes in magnitudes and phases of the spectral components. There is no direct relationship to autoregressive all-pole filters, as there was for the Burg analysis of power spectra. The significant advantage to the Boltzmann ME data processing is that the time points need not be uniformly spaced in time-frequency analysis and the spatial array sensors need not be uniformly spaced in spatial-wavenumber analysis. The spatial sensors need not even be colinearly located, but can be arranged in two or three spatial dimensions to enhance the reception of signals from preferred directions. A second significant advantage to using Boltzmann rather than Gibbs-Burg ME data processing is the capacity of the former to handle more vastly underdetermined problems.

The Boltzmann data processing obtains a ME spectrum of a process more directly than does the Gibbs-Burg processing, according to J. Skilling and S. F. Gull [10], who quite correctly say,

> The Burg form $\sum \log p$ is a measure of the entropy of the probability distribution function (p.d.f.) of the time series, whereas spectrum estimation is mostly an image reconstruction problem in which the entropy ought to be that of the spectrum profile itself.

An algorithm is described for converging by iteration on either a discrete frequency or discrete wavenumber amplitude spectrum, and is illustrated by examples. These show the Boltzmann method compares favorably with the Gibbs-Burg method of Chap. 5 and the number of well-resolved spectral lines can be many times the number allowed by conventional Fourier means.

A generalized-inverse expression for the ME amplitude spectrum indicates that the Boltzmann solution *always exists*. There is no ill behavior when the number of constraints is $\gtrsim$ half the number of unknown amplitudes, as there was for the Boltzmann geotomographic method of Chap. 3.

A disadvantage of the Boltzmann approach is that it appears to be more sensitive to noise than the Gibbsian method. As a rule of thumb, the Boltzmann method will satisfactorily superresolve spectra when the signal-to-noise ratio is 10 dB or greater (signal field/noise field ≥ 3.2), compared to the superresolution reported by Marple [9] and quoted in Chap. 5 for SNR ratios of < 0 dB!

Chapter 7 is devoted to a summary of the ME photon image restoration analysis published by R. Kikuchi and B. H. Soffer [11]. Their very detailed development of the statistics of photon emission and detection yields a general expression for the maximum entropy of the number of photons emitted from an image pixel within a narrow frequency range during the measurement time, as a function of the average emission rate and the number of degrees of freedom, z. This expression takes essentially one of two forms for radio astronomy work, the "log B" form or the "B log B" form, B being pixel brightness. In many—but certainly not all—situations z is so much larger than the average emission rate of each pixel that the entropy in the object objective function can be written in the Shannon form $-\sum_j B_j \log(B_j/B_T)$. This objective function is made an extremum for all variations δB_i, yielding a set of equations for the B_j/B_T in terms of the Lagrange multipliers. These are evaluated to satisfy the constraints.

S. J. Wernecke and L. R. D'Addario's work [12] is summarized as an example of data processing the "log B" form of the entropy, so that the ME spectra deviate from their measured values no more than the noise standard deviation.

In the final Chap. 8, I first show how to use the ME-Boltzmann method to deduce the probability of failure of an ensemble of one type of system component under stress, from incomplete manufacturer's data. Then, I develop the Boltzmann and Gibbs formulas for failure probabilities of independent components from limited path-failure data in a large system. The system is represented as a network of branches (paths) interconnected to nodes (components). Accurate values for these component failure probabilities enable computation of all the path failure probabilities and the probability of system failure.

Only the ME-Gibbs method of estimating component nonfailure probabilities is useful because rather tight bounds must be imposed on them. A synthetic-data example illustrates this point.

I acknowledge with pleasure and gratitude the many stimulating discussions with Dr. B. H. Soffer of Hughes Research Laboratories, Malibu, CA, about the fascinating subject of entropy.

Robert M. Bevensee
BS, MS, EE, PhD (Stanford University)

REFERENCES

[1] J. Skilling and S.F. Gull, Algorithms and applications, In *Maximum Entropy and Bayesian Methods in Inverse Problems,* C. Ray Smith and W.T. Grandy (eds.). Dordrecht, Netherlands: D. Reidel, 1985, p. 93.

[2] E.T. Jaynes, On the rationale of maximum-entropy methods, Proc. IEEE, *79*, September, 1982. pp. 939–952.

[3] E.T. Jaynes, Prior probabilities, IEEE Trans. Sys. Sci. Cyber., *SSC-4,* September, 1968, pp. 227–241.

[4] B.R. Frieden, Dice, entropy and likelihood, Proc. IEEE, *73,* December, 1985. pp. 1764–1770.

[5] F.W. Sears, *Thermodynamics,* Cambridge, MA: Addison-Wesley, 1950.

[6] J.P. Burg, Maximum entropy spectral analysis, Proc. 37th Meeting of the Society of Exploration Geophysicists, 1967. Reprinted in *Modern Spectrum Analysis,* D.G. Childers (ed.). IEEE Press, 1978, pp. 34–41.

[7] R.N. McDonough, Maximum-entropy spatial processing of array data, *Geophys. 39,* Dec, 1974, pp. 843–851. Reprinted in *Modern Spectrum Analysis,* D.G. Childers (ed.). IEEE Press, 1978, pp. 72–80.

[8] E. Rietsch, The maximum entropy approach to inverse problems, *Jour. Geophys., 42,* 1977, pp. 489–506.

[9] S.L. Marple, Frequency resolution of fourier and maximum entropy spectral estimates, *Geophys. 47,* Sept., 1982, pp. 1303–1307.

[10] J. Skilling and S.F. Gull, op. cit., p 84.

[11] R. Kikuchi and B.H. Soffer, Maximum entropy image restoration, I. The entropy expression, *Jour. Opt. Soc. Am., 67,* Dec., 1977, pp. 1656–1665.

[12] S.J. Wernecke and L.R. D'Addario, Maximum entropy image restoration, IEEE Trans. Comp., *C-26,* April, 1977, pp. 351–364. Reprinted in *Modern Spectrum Analysis,* D.G. Childers (ed.). IEEE Press, 1978, pp. 287–300.

SOME NOMENCLATURE

$A^{\angle b}$ denotes a complex number of amplitude or magnitude A and phase b.

$a \to b$ means "a approaches b".

* as superscript denotes complex conjugate, as A^*.

a bar over a symbol denotes a vector of many components, as $\bar{x} = (x_1, x_2, \ldots, x_N)$.

a boldface symbol denotes a matrix, as **M**.

{ } denotes an ensemble or set of items, as $\{x_k\}$ = all $x_1, \ldots, x_N$

β denotes Lagrange multiplier

$\langle\ \rangle$ denotes expectation or average over an ensemble of possible objects.

δ, Δ denotes small changes, as δF, Δx. δ also denotes the Dirac delta function

$\hat{d}$ denotes resolution or discrimination level, as $\hat{d}x$.

det, or $\| \ \|$, determinant of a matrix. As det $\mathbf{M}$ or $\| \mathbf{M} \|$.

dot denotes a vector product in $\overline{A} \cdot \overline{B}$.

e, base of natural logarithms, 2.71828 . . .

$\eta(z)$ denotes a Lagrange-multiplier function of z.

! denotes factorial, as $N! = N(N-1)(N-2) \cdots (2)(1)$.

$>$, greater than, as $A > B$.

h is entropy per element in an object configuration, $-\sum p_i \log(p_i)$, where $p_i = x_i/x_T = n_i/N$ (Shannon form). Also Planck's constant.

H is total entropy, hx_T or hN.

∞, infinity

$\int$, integral

$\hat{}$, caret over a symbol, denotes a current ME estimate, as $\hat{x}_i$. (but see $\hat{d}$)

$<$, less than, as $A < B$

log denotes logarithm to the base $e = 2.71828$ . . . .

$-$ superscript denotes a generalized inverse matrix, as A^-.

-1 superscript denotes the inverse of a square matrix, as $\mathbf{A}^{-1}$, where $\mathbf{A}^{-1}\mathbf{A} = \mathbf{A}\mathbf{A}^{-1} = \mathbf{U}$, the unit matrix.

$| \ |$ denotes magnitude, as $| G |$

O denotes object

∂ denotes partial derivative, as $\partial f/\partial x$

$\prod$ denotes product, as $\prod_i^N a_i = a_1 a_2 \cdots a_N$.

Prob represents probability

ρ, density (gm/cm^3 or kg/m^3 for example)

σ, standard deviation

$\sum$, sum, as $\sum_1^N a_i = a_1 + a_2 + \cdots + a_N$.

t, superscript, denotes matrix transpose, as $\mathbf{M}^t$.

Trace, sum of diagonal elements of a matrix.

1

Philosophical Foundations

1.1 INTRODUCTION

This book follows two general principles to deduce an incompletely defined object (in the general sense) from its image are (1) make as few assumptions as possible and (2) obtain a "most probable" object consistent with the data (image). We derive the three best known distribution functions in physics by the maximum entropy (ME) method with just a few assumptions. Then, we outline the general problem of inverting a noisy image to recover a possibly noisy object with various noise sources and the statistical uncertainty inherent in the data processing itself. We relate the maximum entropy and maximum likelihood methods of inversion to Bayes's Theorem in statistics. Then, we show how to modify the entropy expression to allow for known or prior biases in the probabilities of a parameter distribution for an object.

This brings us to a comparison of the physical entropy of a system with information about that system. The Maxwell demon, first believed able to violate the second law of thermodynamics by steadily decreasing the entropy of a system of molecules, has been shown to suffer a greater entropy increase by virtue of the "information" reaching his eyes than the entropy decrease he is able to effect in the system. The strategy of a gambler with a noisy communication channel is to bet so as to maximize a function—his entropic growth of capital—of the objects (sporting events) subject to images received (outcomes of those events). This strategy makes the increase in his entropic growth equal to the rate of transmission of information over the channel as defined by C. E. Shannon. Then, we turn to a constant-total-money milieu, in which various entities (individuals, banks, corporations) all transact with each other and banks can borrow from the Federal Reserve System, and show that the net change in money entropy, δH, for a large number of transactions fluctuates with time and H may not be near a maximum. But each entity acts to increase its own entropy capital with each transaction. Finally, we give an argument that suggests "the rich get richer and the poor get poorer" because this process increases the entropy of wealth.

1.2 PRINCIPLES OF DATA INVERSION

In this text, we are concerned only with problems which are "underdetermined" in the sense that there are not enough data to solve each problem uniquely. This means, if we look at an object, say the earth, and try to resolve its density distribution by the few data available—its total weight, moment of inertia, and perhaps a few gravitational constants—we are unable to do so uniquely. Our only recourse is to obtain a "best estimate" of that "fuzzy" object according to a mathematical criterion which makes good common sense.

A good "First Principle of Data Reduction" is this

> The result of any transformation imposed on the experimental data [to resolve an object] should incorporate and be consistent with all relevant data [obtained from an "image"] and be maximally noncommittal with regard to unavailable data.
>
> —J. G. Ables, from the Astronomical Astrophysics Supplemental Series, Vol. 15, pp. 383–393, 1974 [1.1]

The maximum entropy methods espoused in this book are consistent with this principle.

In order to avoid "reading more information into the data" than is justified one should also observe that,

> A lack of information cannot be remedied by any mathematical trickery.
>
> —C. Lanczos, *Linear Differential Operators*, 1961 [1.2]

To this we add another tenet of this book

> Mathematical "trickery" [such as maximum entropy data processing] can sometimes replace missing information to yield a most-probable answer.
>
> —R. M. Bevensee, URSI XXI International Symposium, Florence, Italy, 1984 [1.3]

It is very easy to make assumptions, either expressed or implied, about inversion problems of importance in many scientific areas in order to obtain more tractible solutions. In the course of solving various problems by the maximum entropy method, we shall try to minimize the assumptions and state clearly what they are. Then, the reader will appreciate why such a powerful method as the entropy method is required to solve a wide range of underdetermined scientific problems.

Perhaps the best examples of vastly underdetermined problems are those of classical and quantum physics, where one derives measurable properties of astronomically large collections of atoms and molecules from several measured data. These properties are obtained from so-called distribution functions, which give the relative numbers of particles in a unit volume with various energies and momenta. In the next section we derive the three most famous distribution functions of physics: the Maxwell-Boltzmann (M-B) for classical particles, the Fermi-Dirac (F-D) for electrons, and the Bose-Einstein (B-E) for photons or light quanta.

1.3 THERMODYNAMICS AND STATISTICAL MECHANICS

We derive the three prominent distribution functions of physics by the maximum entropy method. Each of these gives the distribution in energy of a vast number of particles in a unit volume—classical gas molecules, electrons, or photons. The derivations follow closely those of F. W. Sears [1.4].

1.3.1 The Maxwell-Boltzmann (M-B) Distribution Function

This function describes the relative number of "particles" in a unit volume (meter^3) of a system with energy between w and w + dw, given the average energy of the particles as a datum. The M-B function is appropriate for classical or nonquantum particles such as gas atoms or molecules with negligible interactions among them.

There is an important philosophical point to be made here which concerns every problem solved with the ME method. We are neglecting the mutual interactions of particles in our derivation of the M-B function. This means either they are truly negligible in a given problem or they are not and we are ignorant about them. In the first case the M-B function will characterize the true system with least bias and enable more accurate prediction of measured parameters such as pressure. In the second case, the function will furnish the most reasonable description of the system subject to more available information. If still more information about the interactions becomes available we can proceed to incorporate it into the derivation of a "modified" M-B function. This will furnish a "sharper" description of the system. The point is that the ME method does not require us to make unwarranted guesses about the behavior of the system under study.

Of course, if we find subsequently that the M-B function does not enable us to predict the measured behavior of a system we should proceed to incorporate some constraints into the ME framework to yield a more realistic; that is, predictive, distribution function. This, in fact, happened historically: electrons were found to have a different heat capacity than predicted by the M-B function, and this forced physicists to incorporate an Exclusion Principle for their statistical behavior. The result was the Fermi-Dirac distribution function, to be derived in Sec. 1.3.2.

Returning now to our system of noninteracting particles, we do assume they are distinguishable, and all of mass m. This implies we can label all of them within the unit volume under study and assign them to different energy groups. We focus on a huge number N of particles and picture, in Fig. 1.1, a cell labeled i in energy space, with center at w_i and width Δw_i. This cell is actually a spherical shell in velocity space, of average radius v_i and width

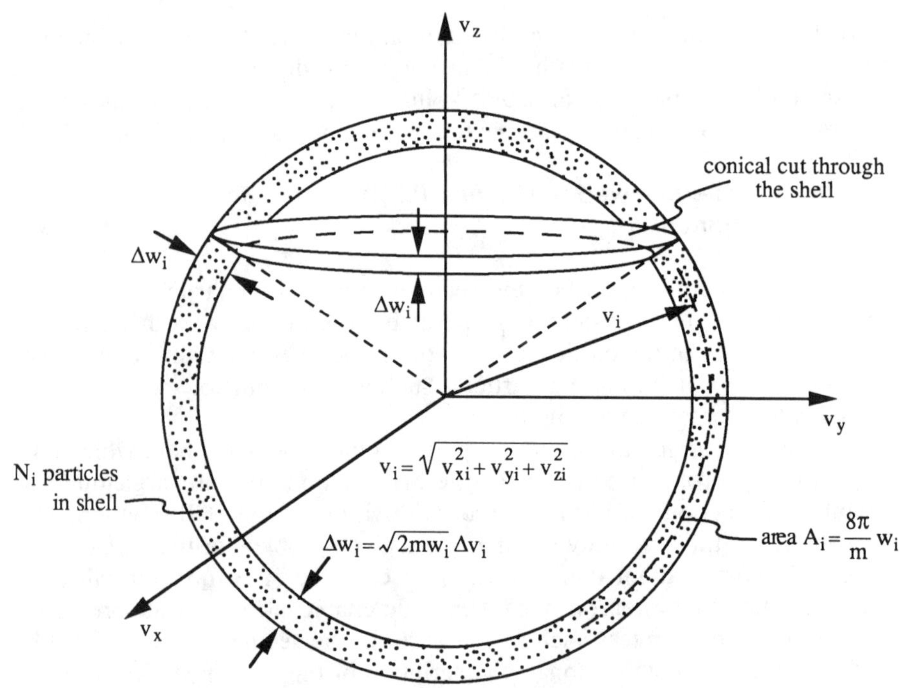

Figure 1.1 A cell i in energy space of average energy w_i and width Δw_i, containing N_i particles (dotted), each of mass m. The cell is a shell in velocity space of radius $v_i = \sqrt{2w_i/m}$ and width $\Delta v_i = \Delta w_i/mv_i$.

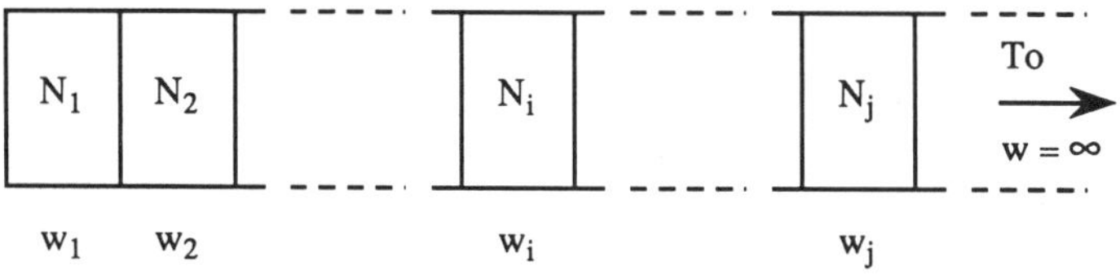

Figure 1.2 Cells of particle energy. The ith cell (or shell) of Fig. 1.1 of average energy w_i contains N_i particles.

Δv_i, obtained from the kinetic energy expression (neglecting gravitational energy)

$$w_i = \frac{1}{2} m v_i^2, \qquad \Delta w_i = m v_i \, \Delta v_i \tag{1.1}$$

$$v_i^2 = v_{xi}^2 + v_{yi}^2 + v_{zi}^2, \qquad \text{constant across the shell} \tag{1.2}$$

We specify N_i particles in the cell at any time and neglect any fluctuations of N_i with time. We seek that distribution function $f(w_i)$ which gives the number of particles $N_i = Nf(w_i)\Delta w_i$ within the shell of Fig. 1.1. For mathematical convenience we treat the energy scale as discrete, as shown in Fig. 1.2. We now seek a most probable distribution of the numbers N_i among the boxes of Fig. 1.2, subject to the constraint information that the total number of particles is $N = \sum N_i$ and that the total energy $W = \sum w_i N_i$ is fixed. We could impose the constraint that $w_{max} < W$ but will not do so for simplicity. The M-B function $f(w)$ will indicate that negligible error is introduced for most problems by allowing a few particles to have energies > W.

The ME method allows us to discard the constraint $w_{max} < W$ at the expense of a less precise $f(w)$ function.

That most probable distribution or arrangement of the N_i in Fig. 1.2 is that arrangement which can be realized *in the greatest number of distinct ways*. The entropy function measures the number of ways of constructing an arrangement of the N_i, so we maximize this entropy function. A maximum entropy solution of any problem treated in this text implies a solution which can be realized in the greatest number of ways, subject to constraints.

We now require a formula which will tell us how many ways there are of putting N_1 distinct particles into box 1 of Fig. 1.2, N_2 particles into box 2, and so on. This formula is precisely the number of combinations C of that arrangement of the N's.

Let us clarify this matter of distinct arrangements by an example. Suppose we have just two energy states, w_1 and w_2 and $N = 4$ particles, and we require the number of combinations C with $N_1 = 3$ in the first state and $N_2 = 1$ in the second. In statistical mechanics, the numbers N_1 and N_2 define a *macrostate*. Labeling the distinguishable particles by letters a, b,

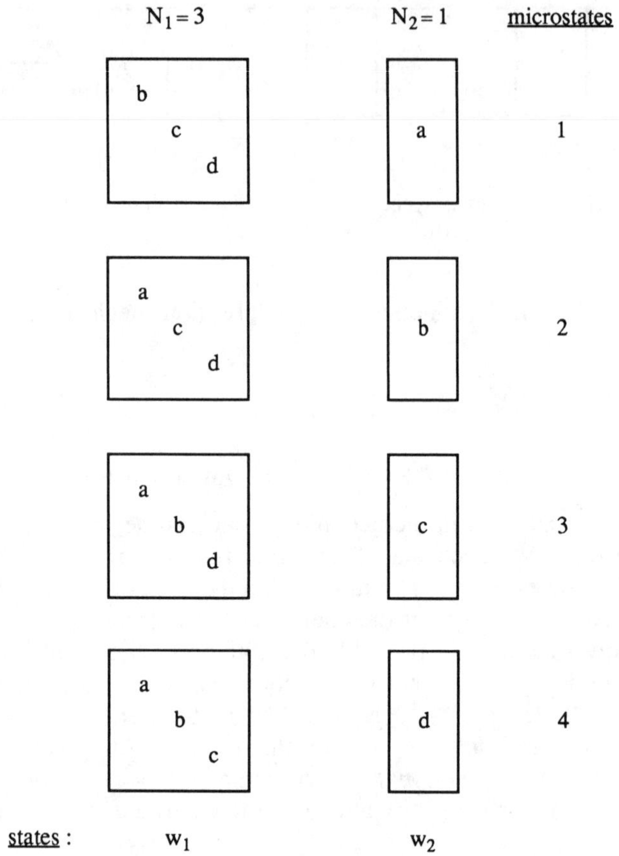

Figure 1.3 There are four distinct ways (microstates) of having the arrangement (macrostate) $N_1 = 3$ and $N_2 = 1$ of the four particles $N = 4$ among the two energy states w_1 and w_2.

c, d we discover there are only four distinct ways—called *microstates*— of realizing this macrostate, as shown in Fig. 1.3. Microstate 1 has the three particles b, c, and d in state w_1, but the order of these is immaterial.

The formula C for the total number of distinct microstates or combinations of microstates with N_1 particles in state $w_1, \ldots, N_i$ in state $w_i, \ldots,$ is

$$C = \frac{N!}{N_1!N_2! \cdots N_i! \cdots} = \frac{N!}{\Pi\, N_i!}, \qquad N = \sum N_i \tag{1.3}$$

where n! is read as n(n − 1)(n − 2) . . . (2)1, and the denominator is the product of all the N_i!-factorials. (0! is defined as 1). Applying this formula to Fig. 1.3, we obtain C = 4!/(3!1!) = 4.

The most probable, maximum-entropy, macrostate of the N_i is the one which maximizes C or log(C) (all logarithms taken to the base $e = 2.71828 \ldots$) because log(C) increases as C does so. It is considerably simpler in practically all problems to maximize log(C), and so we define the entropy H as log(C)

$$\text{entropy } H = \log(C) \tag{1.4}$$

(In thermodynamics H = klog(C), k being Boltzmann's constant). log(C) is convenient to work with because of Stirling's approximation,

$$n! \approx \sqrt{2\pi n}\,(n/e)^n \left[1 + \frac{1}{12n}\right], \qquad n \gg 1 \tag{1.5}$$

This gives for Eq. (1.3)

$$\begin{aligned} \log C &= \log(N!) - \sum_i \log(N_i!) \\ &\approx \frac{1}{2}\log(2\pi) + \frac{1}{2}\log N + N\log N - N + \log\left(1 + \frac{1}{12N}\right) \\ &\quad - \sum_i \left[\frac{1}{2}\log(2\pi) + \frac{1}{2}\log N_i + N_i \log N_i + \log\left(1 + \frac{1}{12N_i}\right)\right] \end{aligned} \tag{1.6}$$

With $\log(1 + 1/12n) \approx 1/12n$ for $n \gg 1$ we have, ignoring the $\log(2\pi)$-terms which merely shift the origin of the entropy axis, and with $N = \sum N_i$,

$$\begin{aligned} \log C &\approx \left(N + \frac{1}{2}\right)\log N + \frac{1}{12N} - \sum_i \left[\left(N_i + \frac{1}{2}\right)\log N_i + \frac{1}{12N_i}\right] \\ &= N\log N - \sum_i N_i \log N_i + \frac{1}{2}\log\left(\frac{N}{\prod N_i}\right) + \frac{1}{12}\left(\frac{1}{N} - \sum \frac{1}{N_i}\right) \\ &= \left(N + \frac{1}{2}\right)\log N - \sum\left(N_i + \frac{1}{2}\right)\log N_i + \frac{1}{12}\left(\frac{1}{N} - \sum\frac{1}{N_i}\right) \end{aligned} \tag{1.7}$$

In thermodynamics and in all our problems we will treat systems for which each N is $\gg 1$, so for practical purposes the last two terms in Eq. (1.7) are negligible. Then, Eq. (1.7) reads

$$\begin{aligned} H = \log C &\approx N\log N - \sum_i N_i \log N_i \\ &\approx -\sum_i N_i \log\frac{N_i}{N} \end{aligned} \tag{1.8}$$

This form shows how convenient it is to maximize log(C) rather than C to obtain a maximum-entropy distribution of the N_i.

We now proceed straightforwardly to obtain the M-B distribution function by maximizing H of Eq. (1.8) subject to the constraints of $N = \sum N_i$ and $W = \sum N_i w_i$. It proves convenient in general to utilize the method of Lagrange multipliers, as follows. We write the first constraint as

$$\sum_i N_i - N = 0,$$

multiply it by an unknown factor β_1, then write the second constraint as

$$\sum_i w_i N_i - W = 0,$$

multiply it by a factor β_2, and add these expressions to log(C), forming an *objective function* F

$$F = \log C + \beta_1 \left(\sum_i N_i - N \right) + \beta_2 \left(\sum_i w_i N_i - W \right) \tag{1.9}$$

in which N and W are fixed parameters. Maximizing F by varying the N_i will also maximize H = log(C) provided the constraints are maintained so that the parenthetical expressions in Eq. (1.9) do not vary.

To maximize F we vary a typical N_j by δN_j in Eq. (1.9), using Eq. (1.8) for log(C), and set $\delta F = 0$ [which maximizes rather than minimizes F because $\partial^2 F/\partial N_j^2 = -1/N_j < 0$ from Eq. (1.9)]. This results in

$$0 = -\left[\delta N_j + \delta N_j \log\left(\frac{N_j}{N}\right) \right] + \beta_1 \delta N_j + \beta_2 w_j \delta N_j \tag{1.10}$$

The δN_j-factor cancels and we get

$$\log\left(\frac{N_j}{N}\right) = -1 + \beta_1 + \beta_2 w_j \tag{1.11}$$

To cast this into a more convenient form we replace β_1 and β_2 by the new parameters A and β defined as: $\log(A) = -1 + \beta_1$, $\beta_2 = -\beta$. Then, Eq. (1.11) becomes the Maxwell-Boltzmann or M-B distribution function

$$N_j = NAe^{-\beta w_j}, \qquad w_j = \frac{1}{2} m(v_{xj}^2 + v_{yj}^2 + v_{zj}^2) \tag{1.12}$$

M-B distribution function. In practice this is considered a continuous distribution over $0 < w < \infty$, and the fractional number, f, of particles per unit volume of energy or velocity space in Fig. 1.1 is

$$f(w) = Ae^{-\beta w}, \qquad w = \frac{1}{2} m\,(v_x^2 + v_y^2 + v_z^2) \tag{1.13}$$

Parameters A and β are determined by the constraints in Eq. (1.9), which are, for the continuous distribution,

$$\int f(w)dw = \frac{A}{\beta} = 1 \tag{1.14}$$

and

$$\int wf(w)dw = \frac{A}{\beta^2} = \langle w \rangle \tag{1.15}$$

Here $1/\beta = \langle w \rangle = w/N$ is the known average particle energy. β is evaluated in classical physics as 1/kT, k being Boltzmann's constant and T the absolute temperature.

To recapitulate, we have derived the classical M-B distribution function for particles in energy, or velocity, space with just the two constraints of a specified number of particles N and their total energy W. We obtained the most probable or maximum-entropy arrangement, or macrostate, of the N by counting the total number of microstates in a given macrostate and maximizing the logarithm of the combinatorial formula for this number (ie., the entropy). We assumed the particles were distinguishable.

The derivation of Eq. (1.12) from F of Eq. (1.9) was only possible because the entropy $H = \log(C)$ is a nonlinear function of the N_i. If H had been of linear form $\sum a_i N_i$ the variational procedure would not have yielded a formula for N_i. This illustrates the general principle for the solution of an underdetermined problem with *linear* constraints: the function to be extremized must be a *nonlinear* function of the object parameters.

In the next section, we will repeat the reasoning but for indistinguishable particles (electrons), which obey an "Exclusion Principle" in action (momentum x position) space.

1.3.2 The Fermi-Dirac (F-D) Distribution Function (for Electrons)

Like the M-B function of Sec. 1.3.1, the F-D function tells us the relative number of electrons in an energy range w to $w + \Delta w$. But the F-D function computes the number of available states in the range Δw according to the quantum properties of electrons. The electron states are counted in discrete units of momentum-position or action space, and no more than two electrons with opposing spins may occupy an elemental cell of volume h^3(h = Planck's constant) in this space. This expresses the Pauli Exclusion Principle.

A small cell in action space is depicted in Fig. 1.4. The ith cell in a small volume $\Delta x \Delta y \Delta z$ of physical space and $\Delta p_x \Delta p_y \Delta p_z$ of momentum space, centered at energy $w_i = p_i^2/2m$, with $p_i^2 = p_{xi}^2 + p_{yi}^2 + p_{zi}^2$, is shown. It contains N_i electrons among the n available states.

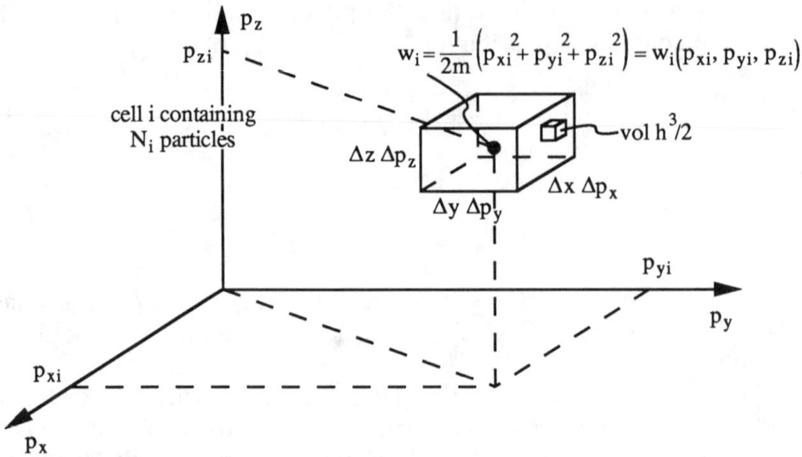

Figure 1.4 A cell of volume dV $= \Delta p_x \Delta p_y \Delta p_z \Delta x \Delta y \Delta z$ in six-dimensional action space. N_i particles reside among the $n = 2dV/h^3$ available states in the cell. The average energy w_i of these states is a function of p_{xi}, p_{yi}, and p_{zi}.

The Uncertainty Principle determines this number n by declaring we cannot resolve electrons any closer in action space than $\Delta p_x \Delta x = \Delta p_y \Delta y = \Delta p_z \Delta z = h$. Therefore, the number of distinguishable states for electrons in the cell of Fig. 1.4 is $\Delta p_x \Delta x \Delta p_y \Delta y \Delta p_z \Delta z / h^3$. By the Exclusion Principle we can have at most two electrons of opposite spins in each state, so the number of subcells n containing at most one electron is

$$n = \frac{\Delta p_x \Delta x \Delta p_y \Delta y \Delta p_z \Delta z}{h^3/2} = \frac{2\Delta p_x \Delta x \Delta p_y \Delta y \Delta p_z \Delta z}{h^3} \tag{1.16}$$

To derive the F-D distribution function, we arrange our cells along an energy scale as in Fig. 1.2. Figure 1.5 shows the new arrangement, with cell i resolved into n available states or subcells, each of volume $h^3/2$. Each subcell can hold at most only one of the N_i electrons in the w_i-cell. To obtain a most probable, maximum-entropy arrangement, or macrostate, of the N_i—that arrangement which can be constructed with the greatest number of distinguishable microstates—we must write an expression for the number of combinations C of microstates which will realize a given macrostate subject to the conditions that (a) we cannot distinguish individual electrons, and (b) no more than one electron can fall in any subcell of Fig. 1.5.

The formula for C is obtained by first focussing attention on a typical cell-say the kth. The number of distinct ways of spreading N_k indistinguishable electrons among n cells is precisely the number of combinations obtained by dividing the n cells into two groups, with N_k (occupied) in one group and $(n\text{-}N_k)$ unoccupied in the other. This is

$$\frac{n!}{N_k!(n - N_k)!} \tag{1.17}$$

For example, the reader may verify that the total number of distinct arrangements of three electrons among $n = 5$ cells is $5!/(3!2!) = 10$. The number of combinations in each N_i-cell is independent of those in the others. Hence, the total number of combinations or microstates of electrons which realize the $N_1, N_2, \ldots, N_i, \ldots$ macrostate of Fig. 1.5 is

$$C = \prod_i \frac{n!}{N_i!(n - N_i)!}, \qquad N = \sum N_i \tag{1.18}$$

This replaces the M-B expression Eq. (1.3).

We now proceed directly to simplify the entropy, log(C), and form the objective function F of Eq. (1.9), with the same constraints of $\sum N_i = N$ and $\sum w_i N_i = W$ as before. With Eq. (1.5) we readily obtain

$$\log C \approx n(\log n - 1) - \sum_i [N_i(\log N_i - 1) + (n - N_i)\log(n - N_i) - (n - N_i)] \tag{1.19}$$

After substituting this into Eq. (1.9) and setting $\delta F = 0$ for arbitrary δN_i,

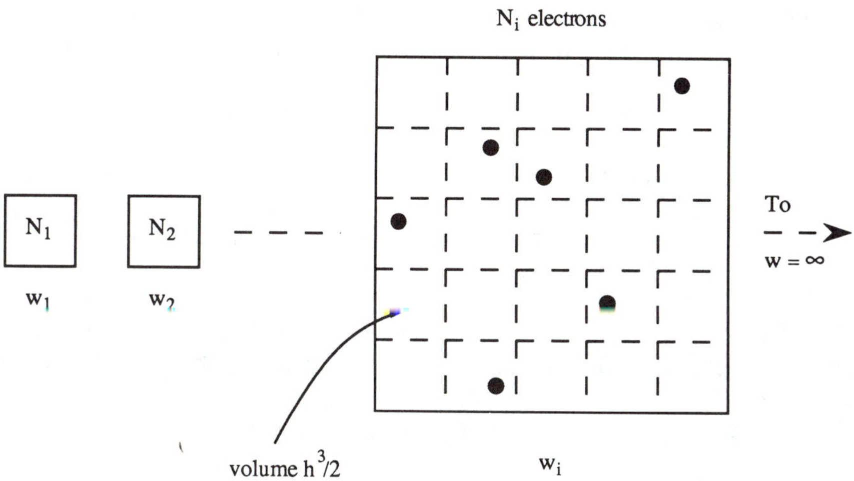

Figure 1.5 Energy cells of electrons for the F-D statistics. The ith cell of energy w_i is of size $\Delta p_x \Delta x \Delta p_y \Delta y \Delta p_z \Delta z$ and has n subcells, each of volume $h^3/2$. Each subcell can have only one of the N_i electrons belonging to the w_i cell. The indistinguishable electrons are shown as dots.

we obtain

$$-[\log N_i - \log(n - N_i) - 1] + \beta_1 + \beta_2 w_i = 0 \qquad (1.20a)$$

or

$$\log\left(\frac{N_i}{n - N_i}\right) = 1 + \beta_1 + \beta_2 w_i \qquad (1.20b)$$

The Lagrange multipliers $\beta_{1,2}$ are as yet unknown, so we are free to replace $1 + \beta_1$ by B and β_2 by $-\beta$, whereupon the F-D distribution is cast into the standard form

$$\frac{N_i}{n} = \frac{1}{Be^{\beta w_i} + 1} \qquad (1.21)$$

F-D distribution function. Note that N_i/n is the fraction of the number n of subcells in action space $\Delta p_x \Delta x \Delta p_y \Delta y \Delta p_z \Delta z$, of average energy w_i, which are actually occupied by electrons. $\beta = 1/kT$ as before.

In some situations, notably when counting electron sites available in solids as a function of energy, the number of subcells n of Eq. (1.16) available in a cell of Fig. 1.4 with energy w_i is a known function of i. In such a situation, n becomes n_i in Eqs. (1.17) to (1.19) and the first term in Eq. (1.19) is summed over i. The result is a replacement of n by n_i in the Fermi-Dirac distribution Eq. (1.21).

The F-D distribution Eq. (1.21) approaches the M-B distribution Eq. (1.12) in the classical limit $h \rightarrow 0$ and $n \rightarrow \infty$, such that N_i/n is a very small number in Eq. (1.21). Then, $B\exp(\beta w_i) \gg 1$ and N_i/n has the form of N_j/N of Eq. (1.12), both representing the relative number of electrons occupying a number of available states of average energy w_i.

The conditions of electron indistinguishability and limitation of one electron at most per subcell in the energy cells of Fig. 1.5 have affected profoundly the maximum-entropy distribution function, Eq. (1.21).

1.3.3 The Bose-Einstein (B-E) Distribution Function (for Photons)

We now derive the B-E function in energy or frequency of an assemblage of photons, or light quanta, which resemble classical particles in not being restricted in action space by the Exclusion Principle but resemble electrons by obeying the Uncertainty Principle. Figure 1.4 also describes N_i photons residing in a cell of size $\Delta p_x \Delta x \Delta p_y \Delta y \Delta p_z \Delta z$. According to the Uncertainty Principle, the number of distinguishable *states* for photons in a typical N_i cell is the same as for electrons. But, because we can have at most two photons of opposite polarization in each state, the number of subcells n containing at most one photon is again given by Eq. (1.16).

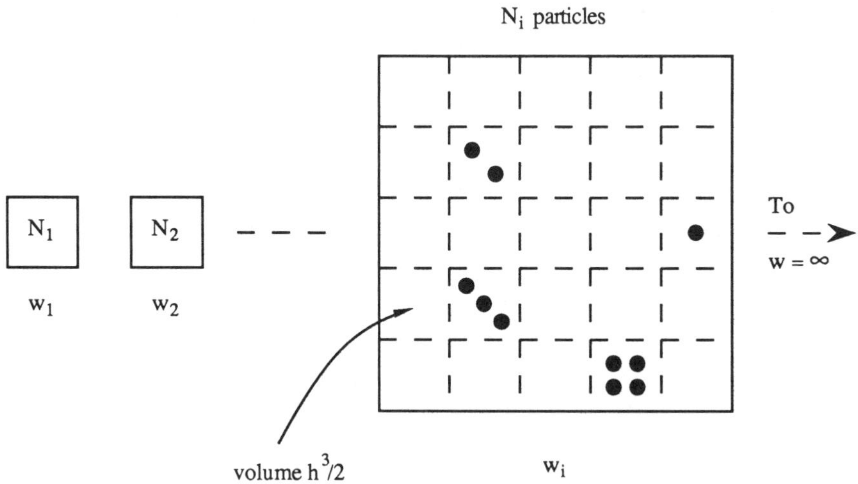

Figure 1.6 Energy cells of photons for the B-E statistics. The ith cell of energy w_i is of size $\Delta p_x \Delta x \Delta p_y \Delta y \Delta p_z \Delta z$ and has n subcells, each of volume $h^3/2$. Each subcell can have any number of the N_i indistinguishable photons (dots) belonging to the w_i cell.

The energy-cell description for photons, from which we will derive the B-E function, has the same subcell structure as in Fig. 1.5 but the number of photons in each subcell of volume $h^3/2$ is unrestricted. So we replace Fig. 1.5 by Fig. 1.6 and show a portion of a possible distribution of the N_i photons in cell i. In order to count the number of combinations C of distinct microstates for a given arrangement or macrostate of all the N_i, we again focus attention on one cell, the kth, in Fig. 1.6. We state without proof, the formula for the number of distinct combinations of microstates of the N_k particles among the n subcells

$$\frac{(n + N_k - 1)!}{(n - 1)!N_k!} \tag{1.22}$$

As an example, let $n = 3$, $N_k = 2$ in the cell of Fig. 1.7, which shows the $4!/(2!2!) = 6$ distinct microstates of the $N_k = 2$ photons. Equation (1.21) is

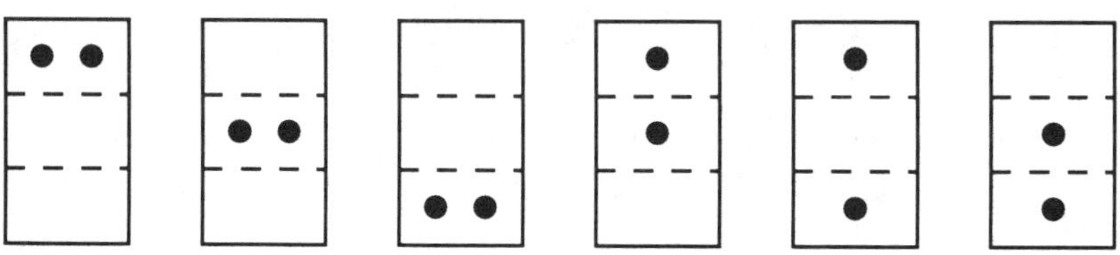

Figure 1.7 The six distinct photon microstates belonging to the $N_k = 2$ macrostate for the kth energy cell of $n = 3$ subcells, each of volume $h^3/2$.

independent of the number of corresponding combinations of microstates in any other cell. Hence, the total number of combinations or microstates of photons which realize the $N_1, N_2, \ldots, N_i, \ldots$ macrostate of Fig. 1.6 is

$$C = \prod_i \frac{(n + N_i - 1)!}{(n - 1)!N_i!} \tag{1.23}$$

This replaces the M-B expression Eq. (1.3).

Proceeding as for electrons, we utilize Eq. (1.5) to obtain

$$\log C \approx \sum_i \Big[(n + N_i - 1)[\log(n + N_i - 1) - 1] - (n - 1)[\log(n - 1) - 1] - N_i(\log N_i - 1) \Big] \tag{1.24}$$

Upon substituting this into the objective function F of Eq. (1.9) and setting variation $\delta F = 0$ for an arbitrary variation δN_i we obtain

$$\log\left(\frac{n + N_i - 1}{N_i}\right) = \beta_1 + \beta_2 w_i \tag{1.25}$$

Now we express β_1 as $-\log B$ and β_2 as $-\beta$ to obtain the B-E distribution function for photons in the standard form

$$\frac{N_i}{n - 1} \approx \frac{N_i}{n} = \frac{1}{Be^{\beta w_i} - 1} \tag{1.26}$$

B-E distribution function. The number n of subcells in each w_i-cell of Fig. 1.6 is assumed $\gg 1$. N_i/n is the fraction of the number of subcells, each of volume $h^3/2$, occupied by photons.

This maximum-entropy B-E distribution Eq. (1.26) predicts experimentally observed properties of a photon "gas" (container of radiant energy) if B is set to 1 and $\beta = 1/kT$, as before. We now derive the Planck formula for the density of radiant energy in a unit volume of an enclosure with walls at temperature T. We use the fact that the energy of a photon at frequency f is $w = hf$ and momentum p is w/c. Then, n of Eq. (1.16) for a unit volume $\Delta x \Delta y \Delta z = 1$ in a small interval of p at $w = w_i$ is

$$n = \frac{2dp_x dp_y dp_z}{h^3} \tag{1.27}$$

Writing $N_i = dN$, the number of photons occupying this six-dimensional element of phase space $dp_x dp_y dp_z \Delta x \Delta y \Delta z$, we have from the B-E function Eq. (1.26)

$$dN = \frac{2dp_x dp_y dp_z}{h^3} \frac{1}{\exp(hf/kT) - 1}, \qquad w = hf = pc \tag{1.28}$$

with

$$p = \sqrt{p_x^2 + p_y^2 + p_z^2} = \frac{hf}{c} \tag{1.29}$$

This depends only on p, so the number of photons in a spherical shell of mean radius p and thickness dp, volume $4\pi p^2 dp$, is

$$d\mathrm{N}_p = \frac{8\pi p^2 dp}{h^3} \frac{1}{\exp(hf/\mathrm{kT}) - 1} \tag{1.30}$$

Finally, we convert this into a distribution in frequency by substituting $p = hf/c$

$$d\mathrm{N}_f = \frac{8\pi f^2}{c^3} \frac{1}{\exp(hf/\mathrm{kT}) - 1} df \tag{1.31}$$

The radiant energy density is hf times this

$$dw_f = \frac{8\pi h}{c^3} \frac{f^3}{\exp(hf/\mathrm{kT}) - 1} df \tag{1.32}$$

Planck distribution function. This is the well-known Planck formula for a thermal distribution of radiant energy.

In retrospect, it is important to observe that we assumed that n, the number of phase space subcells within a unit spatial volume $\Delta x \Delta y \Delta z = 1$ and momentum volume $dp_x dp_y dp_z$ of interest, was $\gg 1$. This is not always true. In Sec. 7.3 we will treat photon image enhancement where n will assume the significance of the number of degrees of freedom of a photon beam on an object pixel relative to an image pixel of the detector. There, n could be $\lesssim 1$ in which case the B-E distribution function would not be appropriate.

1.4 TRANSFORMATIONS BETWEEN OBJECT AND IMAGE

Any data processing scheme to solve the inverse problem of characterizing a system (object) from measurements of its image includes two sources of error, if we neglect data processing numerical error itself (almost always negligible). One is the uncertainty in the object arising from the limited measurement data, exclusive of any noise. We call this *statistical uncertainty*. The other source of error is *noise*, and it can have three components: object noise or fluctuations, transmission noise, and measurement noise. We shall lump these together and call their total *image noise* [1.5].

Figure 1.8 illustrates the general inversion problem of reconstructing an object $\bar{x}^{ob}$ perturbed by n^{ob} from its noisy image $\overline{\mathrm{T}} + \overline{\mathrm{T}}^n$. The bar indicates a quantity which is a vector; that is, has many components. The superscript

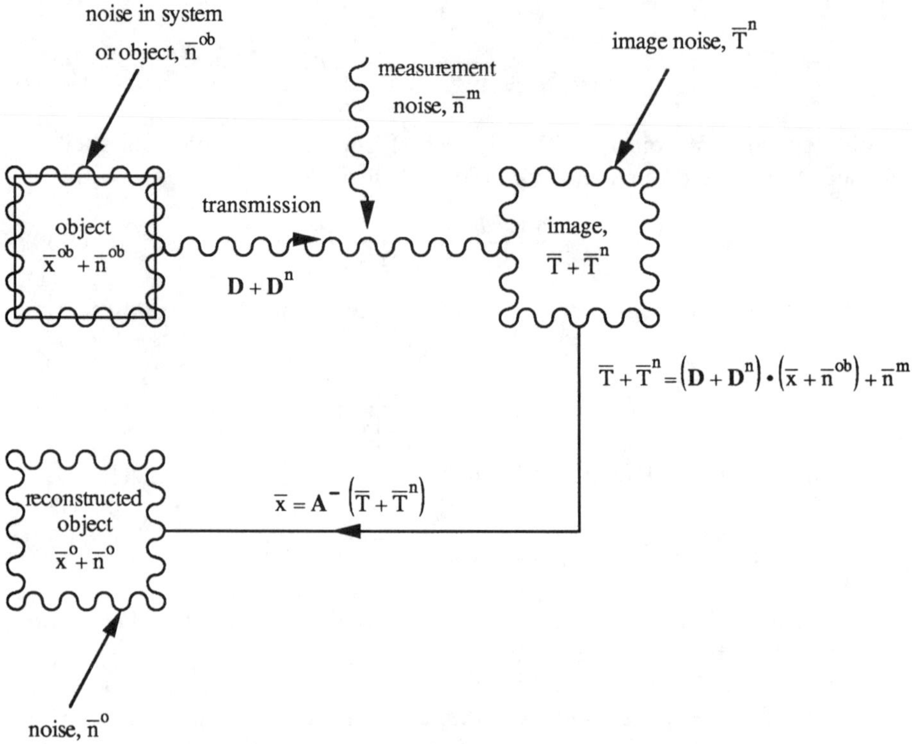

Figure 1.8 The general inversion problem. Noise is indicated by wriggly lines. Noisy object ($\bar{x}^{ob} + \bar{n}^{ob}$) is measured through a noisy transmission medium ($\mathbf{D} + \mathbf{D}^{n}$), producing a noisy image ($\bar{T} + \bar{T}^{n}$). This is inverted by the operation $\mathbf{A}^{-}$ to produce the noisy reconstructed object ($\bar{x}^{o} + \bar{n}^{o}$).

denotes the type of vector. The data processing scheme utilizes a clever inverse operator $\mathbf{A}^{-}$ to transform $\bar{T} + \bar{T}^{n}$ into the noisy reconstructed object $\bar{x}^{o} + \bar{n}^{o}$, "close" to the true object $\bar{x}^{ob}$ in some quantified sense. $\mathbf{A}^{-}$ is the same for $\bar{T}$ as $\bar{T}^{n}$ since we cannot distinguish between a noiseless object of coordinates ($\bar{x} + \bar{n}^{ob}$) and an object $\bar{x}$ perturbed by noise $\bar{n}^{ob}$. Also, the data processing cannot separate the various noise components in $\bar{T}^{n}$; it must invert the entire image ($\bar{T} + \bar{T}^{n}$) to obtain the entire reconstructed object ($\bar{x}^{o} + \bar{n}^{o}$).

Figure 1.8 does not distinguish clearly the statistical uncertainty of $\bar{x}^{o}$ lying in some region near the true $\bar{x}^{ob}$, independent of the perturbing effect of noise, $\bar{n}^{o}$. We show this distinction in Fig. 1.9. Figure 1.9(a) indicates the situation where all noises $\bar{n}^{ob}$, $\bar{\mathbf{D}}^{n}$, and $\bar{n}^{m}$, and hence, $\bar{T}^{n}$, are zero. In Fig. 1.8, $\bar{x}^{o}$ is now denoted $\bar{x}^{ME}$, an ME value. Since there are many possible objects such as $\bar{x}_1$ which could yield the same image $\bar{T}$, there is a range of statistical uncertainty $\delta\bar{x}^{s}$ about $\bar{x}^{ME}$. In Sec. 3.7 we shall discuss the pro-

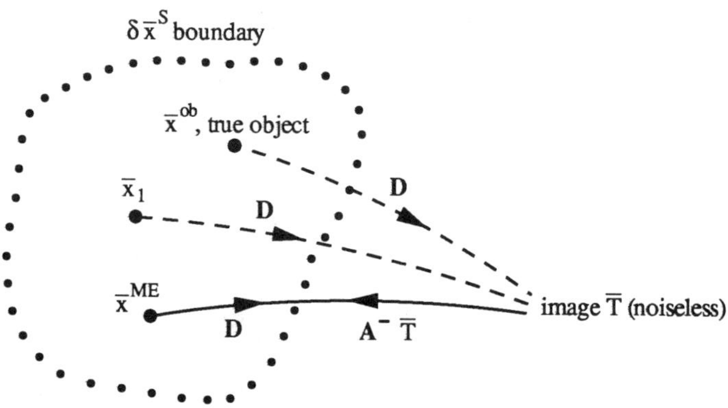

(a) no image noise

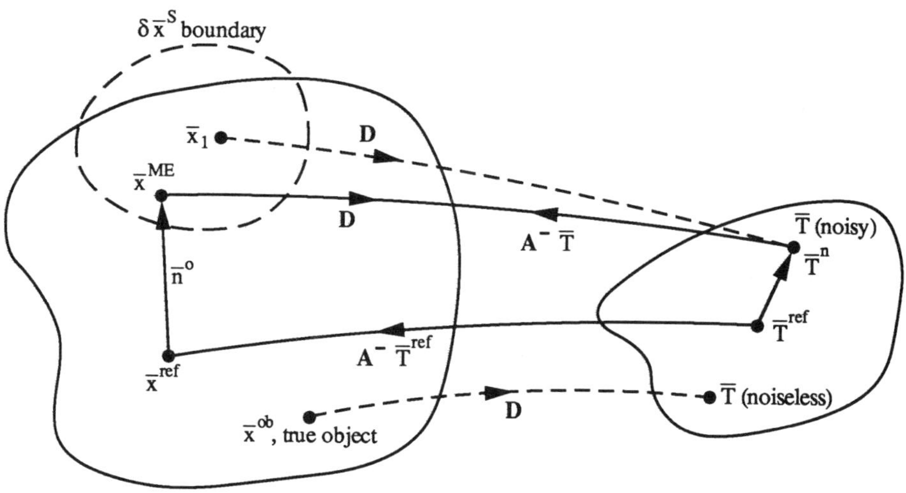

(b) image noise present

Figure 1.9(a) Noiseless situation in Fig. 1.8. True object $\bar{x}^{ob}$ yields an image $\overline{\mathrm{T}}$. The ME inversion operation $\mathbf{A}^-$ independent of noise yields an ME estimate $\bar{x}^{\mathrm{ME}}$. The $\delta\bar{x}^s$-region is a region enclosing a certain percentage of possible $\bar{x}$, like $\bar{x}_1$, which could yield the image $\overline{\mathrm{T}}$. (b) Noisy image ($\bar{n}^{ob}$ and $\mathbf{D}''$ and measurement noise of Fig. 1.8 not shown). Reference image $\overline{\mathrm{T}}^{\mathrm{ref}}$ selected to be close to $\overline{\mathrm{T}}$ in the absence of noise is corrupted by noise $\overline{\mathrm{T}}''$ due to all three noise sources. A noisy $\overline{\mathrm{T}}$ is inverted by $\mathbf{A}^-$ to give $\bar{x}^{\mathrm{ME}}$ perturbed by noise $\bar{n}^0$ from $\bar{x}^{\mathrm{ref}}$. The independent statistical uncertainty $\delta\bar{x}^s$ is superimposed on $\bar{x}^{\mathrm{ref}}$: $\bar{x}_1 = \bar{x}^{\mathrm{ref}} + \bar{n}^o + \delta\bar{x}^s$ is a possible object.

cedure of obtaining *Assurance* intervals for $\delta\bar{x}^s$. An Assurance interval is a specified contour of $\delta\bar{x}^s$ which contains a certain percentage of all possible $\bar{x}$ consistent with the noiseless image $\bar{\mathrm{T}}$. This percentage is precisely the Assurance that the true $\bar{x}$ lies within this contour. One form of Assurance interval for a parameter x_i of $\bar{x}$ is the variance $\langle(\delta x_i^s)^2\rangle$ of a derived distribution function for x_i, the brackets denoting an average over the ensemble of all possible δx_i^s. We shall derive this variance in Sec. 3.7.

In Fig. 1.9(b) we show the general, noisy situation in Fig. 1.7, with the object $\bar{x}$ represented as a point in multidimensional coordinate space. We have not shown the noisy components $\bar{n}^{ob}$, $\bar{\mathbf{D}}^n$, and $\bar{n}^m$, but only the image noise $\bar{\mathrm{T}}^n$ which shifts $\bar{\mathrm{T}}^{ref}$ (a "shrewd" approximation to $\bar{\mathrm{T}}$) to the noisy $\bar{\mathrm{T}}$. The $\mathbf{A}^-$-operation (independent of noise) reconstructs $\bar{x}^{ref}$ from $\bar{\mathrm{T}}^{ref}$ and the reconstructed noisy object $\bar{x}^{ME} = \bar{x}^{ref} + \bar{n}^o$ from noisy $\bar{\mathrm{T}}$. The solid-line contours denote a noisy region of $\bar{\mathrm{T}}$ mapped by $\mathbf{A}^-$ into a noisy region of $\bar{x}^{ME}$. Superimposed on $\bar{x}^{ME}$ we have the dotted Assurance region of $\delta\bar{x}^s$ shown in (a). In Sec. 3.8 we will derive the total variance of an ith component of x, $\langle(x_i - x_i^{ref})^2\rangle$, in terms of the noise autocorrelations $\langle \mathrm{T}_i^n \mathrm{T}_j^n\rangle$ and the statistical $\langle(\delta x_i^s)^2\rangle$.

We distinguish three measurement scenarios with one object $\bar{x}$ in Figs. 1.8 and 1.9, along with their data processing procedures:

1. A single noisy measurement is made, $\bar{\mathrm{T}}^{noisy}$, with no averaging over noise space and no object or transmission noise. Then, image noise $\bar{\mathrm{T}}^n$ = measurement noise $\bar{n}^m$. From Fig. 1.9(b) we have the following relations,

$$\begin{aligned} \bar{x}^{ME} &= \mathbf{A}^-\bar{\mathrm{T}}^{noisy} \\ \bar{x}^{ref}(\bar{\mathrm{T}}^n) &= \mathbf{A}^-(\bar{\mathrm{T}}^{noisy} - \bar{\mathrm{T}}^n) \\ &= \bar{x}^{ME} - \mathbf{A}^-\bar{\mathrm{T}}^n \qquad (\bar{\mathrm{T}}^n \text{ is statistical}) \\ \bar{x}^{tot} &= \bar{x}^{ref}(\bar{\mathrm{T}}^n) + \delta\bar{x}^s = \bar{x}^{ME} - \mathbf{A}^-\bar{\mathrm{T}}^n + \delta\bar{x}^s \end{aligned} \tag{1.33a}$$

This last statement gives a possible $\bar{x}$, $\bar{x}^{tot}$, in terms of a possible noise $\bar{\mathrm{T}}^n$ and possible $\delta\bar{x}^s$. The variance of $\bar{x}^{tot}$ about the $\bar{x}^{ME}$ is, assuming $\bar{\mathrm{T}}^n$ and $\delta\bar{x}^s$ are uncorrelated,

$$\langle|\bar{x}^{tot} - \bar{x}^{ME}|^2\rangle = \langle|\mathbf{A}^-\bar{\mathrm{T}}^n|^2\rangle + \langle|\delta\bar{x}^s|^2\rangle \tag{1.33b}$$

The left side gives a measure of the size of the region around $\bar{x}^{ME} = \mathbf{A}^-\bar{\mathrm{T}}^{noisy}$ within which the true object $\bar{x}$ might lie.

If $\mathbf{A}^-$ adjusts the noise in the solution so that it is not inverse to $\mathbf{D}$ then, Eq. (1.33a) may be modified for one measurement $\bar{\mathrm{T}}^{noisy}$ to read

$$\begin{aligned} \bar{x}^{ME} &= \mathbf{A}^-(\bar{\mathrm{T}}^n)\cdot\bar{\mathrm{T}}^{noisy}, \qquad \text{but} \\ \mathbf{D}\bar{x}^{ME} &= \bar{\mathrm{T}}^{ref} \end{aligned} \tag{1.34}$$

and the noise parameter(s) in $\mathbf{A}^-$ is (are) adjusted until $|\overline{T}^{noisy} - \overline{T}^{ref}|$ is sufficiently small relative to the (known) standard deviations σ_k of the noise in each T_k-parameter of the image.

$$\sum_{k=1}^{N} \frac{|T_k^{noisy} - T_k^{ref}|^2}{\sigma_k^2} \leq N \tag{1.35}$$

One must be careful when employing this scheme not to underestimate the σ_k or noise artifacts may appear in the reconstructed image $\overline{T}^{ref}$.

2. A succession of noisy measurements is made in the presence of stationary noise (time-invariant statistics). These are averaged to obtain $\overline{T}^{ref}$ in Fig. 1.9(b), from which the reconstructed object is $\overline{x}^{ref} = \mathbf{A}^-\overline{T}^{ref}$. A confidence interval for the true $\overline{x}$ is obtained from the statistics of $\overline{x}^{ref} + \delta\overline{x}^s$.
3. Many noisy measurements are made and no information is available about the noise statistics, except that the $\overline{T}^n$ are bounded. One has no choice but to construct an $\overline{x}^{ME}(\overline{T})$ from each noisy $\overline{T}$ and estimate the object $\overline{x}$ from the $\overline{x}$-space distributions of the $\overline{x}^{ME}(\overline{T}) + \delta\overline{x}^s$.

Most ME data processing schemes follow procedure (1) conceptually, neglecting statistical uncertainty. Many procedures accept an ME solution if it predicts an image within a known standard deviation σ of the noise $\overline{T}^n$ according to Eq. (1.35).

If measurement noise is sufficiently high it can preclude the computation of any ME object! This means we cannot find a set of Lagrange multipliers in an objective function like F of Eq. (1.9) by minimizing F with respect to all variations of parameters and solving the resulting set of equations, together with the exact constraint equations including large noise deviations. If this happens, one has no recourse but to accept a collection of less-than-maximum entropy objects which will satisfy the large-noise constraint equations. The many-to-one relationship of possible objects to image usually guarantees that such a collection exists.

An example of this corrupting influence of noise will be mentioned in Sec. 6.7, where a slight amount of measurement noise added to receivers of coherent radiation of objects in the sky (wavenumber resolution) complicates the computation of ME objects. This situation is usually academic, since the assumed coherent nature of the sources allows one to remove the noise by long-time correlation or averaging techniques.

1.5 MAXIMUM ENTROPY INVERSION AND BAYES'S THEOREM

The Bayesian method in statistics is well known for computing the conditional probability of a cause (object here) given the effect (image) from the computed probabilities of the various images given each object. The problem of computing the ME object from its image in Fig. 1.9 is illustrated differently in Fig. 1.10 for the purpose of invoking Bayes's Theorem. The Maximum Likelihood ML object is also indicated; it is found formally in much the same way as the ME object.

The various macrostates of possible objects are represented by rectangles, all the same size to symbolize the same prior probabilities for all the possible objects. Different priors would weight the boxes accordingly. Many possible objects could yield the measured image I_m; for simplicity we

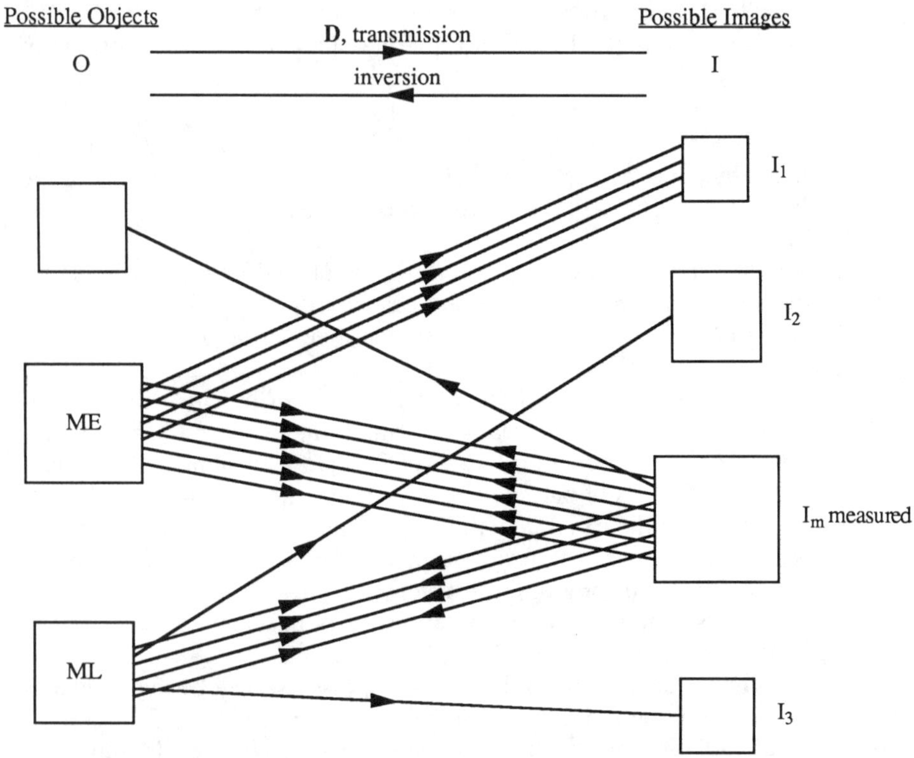

Figure 1.10 Schematic representation of the object-image transformation of Fig. 1.9. Noise can be ignored here; statistical uncertainty causes many objects to correspond to the measured image I_m. The two objects on the left labeled ME and ML are the maximum-entropy and maximum-likelihood objects, respectively. The ME object is the one which can be realized the greatest number of ways. This is indicated by more solid lines from I_m to O^{ME} than any other object, including O^{ML}. The ML object is defined as having the greatest probability of creating the measured image. This is indicated by a greater fraction of lines from O^{ML} to I_m (4/6) than from any other O to I_m. The fraction of lines from O^{ME} to I_m is only 6/10. Prior probabilities would weigh the relative importance of the objects on the left.

show only the ME and ML objects as possible sources for I_m. I_m represents $(\bar{T} + \bar{T}^n)$ in Fig. 1.8 and $(\bar{T}^{ref} + \bar{T}^n)$ in Fig. 1.9(b). The ME object is the one which maximizes $P_{I_m}(O)$—the object probability given the image I_m. Thus, O^{ME} has a greater number of solid lines connecting I_m than does any other object. The ML object is that O^{ML} which maximizes the likelihood $P_O(I_m)$ of the measured image probability. O^{ML} has a greater fraction of its lines to image I_m than does any other object. The subtle difference between the ME and ML data processing schemes will be summarized in the next section.

Measurement noise is implicitly included in the definitions of the various probabilities defined next.

To state Bayes's Theorem [1.6] we define the unconditional probabilities $P(O)$, the so-called prior probability of an object macrostate O. It is often ignored or taken to be a constant for all O within a restricted range (semicomplete ignorance). $P(I) = \sum_O P(O)P_O(I)$, the net probability of an image macrostate I, as computed from all the known conditional probabilities $P_O(I)$. Usually $P(I)$ is treated rather trivially as a normalization constant.

The conditional probabilities are $P_O(I)$ = probability of an image I given a particular object O. These are all computable given the transmission factors D. $P_I(O)$ = probability of an object O given a particular image I.

In terms of these probabilities, Bayes's Theorem states [1.6] that the probability of any possible object O conditioned on the measured image I_m is computed from the "forward" probabilities of image I_m given object O as

$$P_{I_m}(O) = \frac{P(O)P_O(I_m)}{\sum_O P(O)P_O(I_m)} \tag{1.36}$$

Bayes's theorem. The denominator is just $P(I_m)$. $P_{I_m}(O)$ is the relative number of combinations of object-O microstates which realize I_m; $P_O(I_m)$ is the relative number of combinations of image I_m microstates associated with O. Each of these combinations C has the entropy log(C).

Some ME data processing schemes compute an ME prior $P(O)$ for Eq. (1.36) from a smaller problem and maximize $P_{I_m}(O)$ by maximizing the numerator instead, subject to constraints. Most schemes ignore the priors $P(O)$ as being unknown and maximize $P_{I_m}(O)$ directly by maximizing the object entropy subject to constraints.

In general, the ME object O^{ME} is found by maximizing Eq. (1.36) relative to O. If the priors $P(O)$ are all equal they cancel in numerator and denominator of Eq. (1.36) and we have just

$$P_{I_m}(O) = \frac{P_O(I_m)}{\sum_O P_O(I_m)} \qquad [P(O) \text{ all equal}] \tag{1.37}$$

In this case the ME object O^{ME} is found by

$$P_{I_m}(O^{\text{ME}}) = \max_O \frac{P_O(I_m)}{\sum_O P_O(I_m)} \tag{1.38}$$

Here $P_O(I_m)$ is proportional to the number of solid lines from O to I_m in Fig. 1.10, which is proportional to the number of microstates of O which realize measured image I_m. Since the denominator of Eq. (1.37) is independent of O that equation defines O^{ME} as that object with the greatest number of microstates which can realize image I_m.

Now consider the ML object. O^{ML} is obtained, not from Bayes's Theorem, but by maximizing the probability that an object O will give the measured image I_m rather than any other image. Calling this probability $P'_O(I_m)$ we find O^{ML} as

$$P'_{O^{\text{ML}}}(I_m) = \max_O \frac{P(O)P_O(I_m)}{\sum_I P(O)P_O(I)} \tag{1.39a}$$

Since $P(O)$ factors out of the sum it cancels, leaving

$$P'_{O^{\text{ML}}}(I_m) = \max_O \frac{P_O(I_m)}{\sum_I P_O(I)}, \qquad \sum_I P_O(I) = 1 \tag{1.39b}$$

The numerator is proportional to the number of dashed lines from O^{ML} to I_m in Fig. 1.10. This corresponds to the number of solid lines from O^{ML} to I_m. The denominator in Eq. (1.38) is the total number of dashed lines from O^{ML} to the various possible images.

The denominators of Eqs. (1.36), (1.38) and (1.39) are all different, indicating the fact portrayed in Fig. 1.10 that *even if the priors* $P(O)$ *are all treated equal, the ME and ML objects for a given measured image* I_m *will generally differ.*

The next section gives other reasons why the ME- and ML-objects will differ for a given image.

1.6 MAXIMUM ENTROPY VERSUS MAXIMUM LIKELIHOOD

In the preceding section we saw that the maximum entropy object O^{ME} is found by maximizing $P_{I_m}(O)$ with respect to O for the given measured image I_m, which means O^{ME} in Fig. 1.10 is realized the greatest number of ways. Let the object be described by a number of parameters or states, $x_1, \ldots, x_M$. Then, we can write the condition for O^{ME} as

$$P_{I_m}(x_1, \ldots, x_M; O^{\text{ME}}) > P_{I_m}(x_1, \ldots, x_M; O) \qquad \text{for } O \neq O^{\text{ME}} \tag{1.40}$$

Now the statistical Principle of Maximum Likelihood Estimation [1.7] says that if the image is parameterized by $y_1, \ldots, y_N$, and if $P'_O(y_1, \ldots, y_N; I_m)$ is the probability of the measured image I_m given an object O the maximum-likelihood ML object is such that

$$P'_{O^{ML}}(y_1, \ldots, y_N; I_m) > P'_O(y_1, \ldots, y_N; I_m), \qquad O \neq O^{ML} \quad (1.41)$$

Condition Eq. (1.40) corresponds formally to maximization of $P_O(I_m)$, the right side of Eq. (1.38), if the priors $P(O)$ are ignored or all equal in Bayes's Theorem Eq. (1.36). Condition Eq. (1.41) corresponds to maximizing the numerator of Eq. (1.39).

In practice, the two techniques of deducing a "most probable" object are rarely the same because they treat the $\bar{x}$-object parameters differently in Eq. (1.40). An illuminating example of this is the computation of the probabilities of rolling each number on a die as inferred from the outcomes of relatively few rolls [1.8]. In the ME treatment, the probability x_i of rolling the ith number was not introduced at all; rather, the most likely number of times n_i out of N of rolling that number was computed. This served as a *frequency* estimate to x_i. In the ML treatment, the x_i's were introduced as parameters and the most likely values found as parameters of O^{ML}. O^{ME} and O^{ML} were nowhere near each other in their $\bar{x}$-values.

We will not discuss any ML data processing further in this text. However, ML spectral analysis is a powerful method [1.9] which, like ME spectral analysis, adapts to the actual characteristics of interfering noise. Both techniques can be particularly valuable for resolving one or more narrow spectral peaks in frequency or wavenumber [1.10]. For this purpose, the output peaks of the ML spectra tend to reflect the peak powers directly, while those of ME spectra tend to be proportional to the squares of the actual peak powers. Neither method appears to be unduly sensitive to small statistical fluctuations of the estimated correlation functions, which Fourier transforms yield the frequency or wavenumber spectra.

Both the ML and ME methods of frequency or wavenumber spectrum estimation are equivalent to the use of a prediction error filter, which coefficients are obtained by solving the autoregressive equations for the measured autocorrelation function. J.P. Burg has shown [1.11] that the reciprocal of the ML spectrum equals the average of the reciprocals of the ME spectra obtained from the one-port up to the N-port prediction error filters. Here N is the number of sensors in wavenumber-space spectral estimation or the number of time points in frequency-time spectral estimation. This accounts for the generally lower resolution of the ML method. For unequally spaced arrays in the space domain or unequally spaced samples in the time domain, the ME spectrum computation becomes very difficult (by the Burg method), whereas the ML spectrum is still easily calculated.

1.7 THE ENTROPY MODIFIED FOR PRIOR KNOWLEDGE

Suppose we have reason to believe that the probabilities $p_k = x_k/x_T$ of a parameter distribution in an object are close to known "prior" values r_k, with $\sum r_k = 1$, each $r_k \neq 0$. To force the p_k to be close to the r_k in the solution, we modify the entropy by an unknown function G

$$h' = -\sum p_n \log p_n + \mathrm{G}(\bar{p}, \bar{r}) \tag{1.42}$$

We want h' to be an extremum with respect to a variation δp_n when $\bar{p} = \bar{r}$

$$\frac{\partial h'}{\partial p_n} = -(1 + \log p_n) + \frac{\partial \mathrm{G}}{\partial p_n} = 0 \qquad \text{when } \bar{p} = \bar{r} \tag{1.43}$$

A simple non-trivial function that guarantees this is

$$\mathrm{G}(\bar{p}, \bar{r}) = p_n + p_n \log r_n + \mathrm{G}_1(p_1, \ldots, p_{n-1}, p_{n+1}, \ldots, p_\mathrm{N}; \bar{r}) \tag{1.44}$$

Repeating this requirement for other variations yields

$$\mathrm{G} = \sum_{n=1}^{\mathrm{N}} (p_n + p_n \log r_n) = 1 + \sum_n p_n \log r_n \tag{1.45}$$

and this gives us for h'

$$h' = -\sum_n p_n \log \frac{p_n}{r_n} + 1 \tag{1.46}$$

The "1" is a trivial entropy shift, and so we take our modified entropy to include the effect of prior information as

$$h_p = -\sum_n p_n \log \frac{p_n}{r_n} \leq 0 \tag{1.47}$$

h_p has the requisite properties: $\partial^2 h_p/\partial p_k^2 = -1/p_k \leq 0$ for all k, which means h_p has one maximum. At the maximum $\partial h_p/\partial p_k = 0$ subject to $\sum p_k = 1$ yields $\bar{p} = \bar{r}$ and the maximum value of zero in Eq. (1.47).

The solution to a general problem with h_p maximized subject to constraints will place the reconstructed object $\bar{p}$ as close to $\bar{r}$ as possible.

The integral form of Eq. (1.47) with the sign changed,

$$h_c(p, r) = + \int dx\, p(x) \log \frac{p(x)}{\mathrm{r(x)}}, \tag{1.48}$$

goes by the name "cross entropy," and its usefulness is justified by the fact that, given constraint information, there is only one distribution $p(x)$ which can be obtained by a procedure which satisfies the consistency axioms: (1) uniqueness of solution, (2) invariance to choice of coordinate system, (3) system independence, and (4) subset independence [1.12, 1.13]. System independence refers to the description of different independent systems by either their individual probability densities or by one joint density. Subset independence refers to description of one system either by separate conditional densities or by the full system density. This distribution $p(x)$ can be obtained by minimizing cross entropy (or maximizing its negative) subject to the constraints.

The negative of Eq. (1.48) corresponds formally to Jaynes's general entropy expression, in which $r(x)$ is replaced by $m(x)$, an "invariant measure" function in continuous parameter space x (which could be multidimensional) [1.14, 1.15]. If a prior distribution $r(x)$ is known it is the invariant measure function because h_c will be invariant to coordinate transformation [axiom (2)]. If there is no $r(x)$, an $m(x)$ could, in rare instances, be required for coordinate transformation invariance, and would have the effect of forcing $p(x)$ toward a specific form $m(x)$. There is no requirement for a proper $m(x)$ in the case of discrete $\{p_i\}$.

For the discrete case, suppose we specify only L of the p_l as being as close as possible to the r_l, and say nothing about the remaining M of the p_m, L + M = N. The foregoing reasoning applied to the p_l yields

$$G(\bar{p}, \bar{r}) = \sum_{l=1}^{L} (p_l + p_l \log r_l) \tag{1.49}$$

in place of Eq. (1.45). This gives a new entropy h'' of

$$\begin{aligned} h'' &= -\sum_{l=1}^{L} p_l \log p_l + \sum_{l}^{L} (p_l + p_l \log r_l) - \sum_{m}^{M} p_m \log p_m \\ &= -\sum_{l}^{L} p_l \log\left(\frac{p_l}{r_l}\right) + \sum_{l}^{L} p_l - \sum_{m}^{M} p_m \log p_m \end{aligned} \tag{1.50}$$

in place of Eq. (1.46). Again, $\partial^2 h''/\partial p_k^2 = -1/p_k$, for all l and m so h'' has only one maximum. To find it we first set $\partial h''/\partial p_l = 0$ for each of the L p_l and find that each $p_l = r_l$. Then, we set $\partial h''/\partial p_m = 0$ for each m subject to the constraint

$$\sum_{l}^{L} p_l + \sum_{m}^{M} p_m = \sum_{l}^{L} r_l + \sum_{m}^{M} p_m = 1 \tag{1.51}$$

This procedure yields $p_m = A$, a constant, which evaluates to

$$A = \frac{1 - \sum r_l}{M} \tag{1.52}$$

by Eq. (1.51). The maximum entropy is obtained from Eq. (1.50) as

$$h''_{\max} = \sum_{l}^{L} r_l - MA \log A \tag{1.53}$$

The objective function for a problem would be h'' of Eq. (1.50) plus the various constraint relations, each multiplied by an (initially unknown) Lagrange multiplier.

1.8 ENTROPY, INFORMATION, AND ASPIRATION

We use the physicists' definition of entropy in this text, as a measure of disorder. The greater the number of distinct ways of constructing a system the greater is its entropy. Any system in one of its highly ordered states, with few ways of constructing it, is in a low-entropy state. A disordered state can be constructed in more ways, hence has higher entropy. Information, per se, means data or facts. We define "entropic-information" by the reduction in uncertainty of facts which a message achieves. Entropic-information and entropy behave in opposite ways: the fewer the number of ways of constructing a message the lower is its entropy but the greater the reduction in uncertainty upon receipt and the greater is the entropic-information.

We shall now discuss two situations in which entropic-information and entropy have been described as being equivalent in some sense. In both situations one need not invoke the definition of entropic-information at all. In support of this viewpoint we observe that J. W. Gibbs, in his monumental treatise on statistical mechanics [1.16], nowhere mentions the word "information." Then, we discuss an economic situation in which each participant tries to increase his own "money" entropy, but the overall change of that entropy depends on the judgments of all the players. Finally, we sharpen the old adage that "the rich grow richer while the poor grow poorer" on the basis of the tendency of the "entropy of wealth" to increase in time.

1.8.1 The "Maxwell Demon"

The demon was constructed by J. C. Maxwell in an attempt to refute the concept that in any process the net entropy of the universe tends to increase, or, at best, stay constant. The demon stands at a sliding door separating two heat-sealed boxes, one with gas molecules at temperature T_1 and the other with identical molecules at $T_2 > T_1$. There is a Maxwell-Boltzmann distribution of molecules in each box, but more of them travel faster in the hotter T_2 box. When the demon spies a relatively slow (cooler) molecule approaching the door on the T_2 side or a hotter one approaching from the T_1 side it quickly opens the door, lets the molecule pass through, then closes it. After a time the demon has succeeded in cooling the T_1 box and warming the T_2 box. This is contrary to what would happen if the door remained open, and it has apparently decreased the thermodynamic entropy and thereby violated the cherished Second Law of Thermodynamics! [Note: when a "hot" molecule of heat energy dQ passes from the T_1 box to the T_2 box, the net entropy changes by $dQ(-1/T_1 + 1/T_2) < 0$.] Apparently, the visual information gained by the demon has enabled him to reduce the net physical entropy.

Thorough discussion of the situation by L. Brillouin [1.17] and others indicated that the photon energy which furnished the demon's information had been omitted from the entropy balance. The molecular entropy had indeed decreased, but the demon's own entropy increased even more because of the dQ energy of the light quanta entering its eyes. Hence, the net entropy steadily increased while the demons sorted the molecules and the Second Law was saved!

This entire situation can be understood without invoking any measure of entropic-information. One may define a sequence of photons entering the demon's eyes as "entropic-information," equivalent to the negative of a thermodynamic entropy change, but it is not necessary.

1.8.2 The Gambler with Inside Information

This fortunate gambler utilizes information from an available channel to win money and change the "money entropy" of the universe. E. O. Thorp, the "Wizard of Odds," did precisely this for years by counting cards at blackjack [1.18], thereby increasing his money entropy, until the casinos changed the rules in order to increase *their* money entropy. To understand this, assume that when a person receives dS dollars, his or her entropy is increased by dS/S, where S is the capital at the time. This is analogous to the definition of molecular entropy change mentioned earlier. Assuming that Thorp's capital S_T is smaller than a casino's capital S_c at the time he wins dS dollars, the net entropy change is $dS(1/S_T - 1/S_c) > 0$. The casinos try to reverse this trend.

J. L. Kelly explained [1.19] how a gambler could utilize a channel on which a sequence of transmitted symbols (advance outcomes of sporting events) is received, to maximize the expected value of the logarithm of his or her capital. If he or she bets the same proportion of capital on an input symbol to his or her channel each time that symbol is received then his capital will grow exponentially with time. If the odds are consistent with the probabilities of occurrences of the transmitted symbols ("fair" odds, equal to the reciprocals of the probabilities) the maximum value of the exponential rate of capital growth equals the rate of entropic-information flow as defined by C. E. Shannon [1.20]. If the odds are not fair, the maximum exponential rate of growth will be larger than with no channel by an amount equal to this entropic rate of information flow.

If V_N is this gambler's bankroll after N bets, starting from V_0, the mathematical statement is

$$\begin{aligned}\text{expected exponential rate of capital growth} &= \lim_{N\to\infty} \frac{1}{N} \log\left(\frac{V_N}{V_0}\right) \\ &= \text{rate of entropic} - \text{information flow}\end{aligned} \tag{1.54}$$

We have assumed that no "house take" reduces the gambler's winnings.

In this situation, the gambler maximizes a function—the entropic growth of capital—of the "object" (input symbols to his channel), subject to constraints on the bets, by acting on the "image" (received symbols from his or her channel). The strategy could be formulated without recourse to any definition of entropic-information. The equivalences previously mentioned between the entropic rates of capital growth and entropic-information transmission are due to the definitions of entropic-information introduced by Shannon.

1.8.3 The "Money Entropy" of a Group of Investors

Consider a number of investors (including our gambler)—individuals, companies, corporations, and the Federal Reserve System—all trying to increase their profits in buy-and-sell transactions defined in a general sense. We exclude nonprofit corporations and government agencies. The Federal Reserve and member banks try to increase their profits by lending money and collecting it with interest later.

Now label all these entities except the Federal Reserve System by an index i running from one to N, a large number. At a given time let the ith entity have a wealth of S_i and the Federal Reserve have a reserve of S_F it would be willing to lend to member banks, before discouraging further lending with extremely high interest rates.

Assume that the total money supply, S_T, defined as

$$\sum S_i + S_F = S_T \tag{1.55}$$

remains fixed. The entropy of all this capital is

$$H_S = -\sum_i S_i \log \frac{S_i}{S_T} - S_F \log \frac{S_F}{S_T} \tag{1.56}$$

After a large number of transactions in which entity i has increased its capital S_i by δS (negative for a decrease) the change in this entropy is

$$\delta H_S = -\sum_i \delta S_i(\log S_i - \log S_T + 1) - \delta S_F(\log S_F - \log S_T + 1) \tag{1.57}$$

The constancy of the money supply Eq. (1.55) implies

$$\sum_i \delta S_i + \delta S_F = 0 \tag{1.58}$$

and so Eq. (1.57) simplifies to

$$\delta H_S = -\sum_i \delta S_i \log S_i - \delta S_F \log S_F = +\sum_i \delta S_i \log \frac{S_F}{S_i} \tag{1.59}$$

Presumably, the Federal Reserve lending capacity S_F is $\gg$ any one S_i so the logarithm term is positive for each i in the summation.

The net entropy change for a large number of transactions in Eq. (1.59) will not necessarily be positive because the signs of the individual δS_i depend so much on aspiration. An investor i buys a stock for s_i and hopes to sell it later for $s_i' > s_i$, thereby hoping to increase his or her money entropy by

$$\delta H_i = +|s_i| \log\left(\frac{S_i}{S_T}\right) - |s_i'| \log\left(\frac{S_i - |s_i|}{S_T}\right) > (|s_i| - |s_i'|) \log\left(\frac{S_i}{S_T}\right) > 0 \tag{1.60}$$

but may have to unload the stock at a lower price and decrease his or her money entropy instead. Because the profits are not guaranteed for any of these profit-seekers, except the Federal Reserve System—which δS_F does not contribute to Eq. (1.59)—the money supply entropy may fluctuate up or down at random in the course of time.

1.8.4 "The Rich Grow Richer and the Poor Grow Poorer"

This is a popular phrase for a tendency of wealth to equalize among the population according to the ME principle. The situation is represented in Fig. 1.11, in which the vertical axis shows the relative number of people, $p(x)$, with wealth x on the horizontal axis. The solid curve is a plot of the distribution of wealth at time 1 and the dashed curve is a plot at a later time 2. The area under each curve is unity, including all the people, so curve 2 of lower average height is spread out more from the peak. This shows greater extremes of wealth. We now show this is consistent with the principle of increasing entropy.

Let each curve have the bell-shaped normal distribution, according to

$$p(x) = \frac{1}{\sqrt{2\pi}\sigma} \exp \frac{-(x - x_0)^2}{2\sigma^2} \tag{1.61}$$

Equation (1.61) could be derived by modifying the constraints which yielded the M-B distribution function Eq. (1.12) to read: conservation of total num-

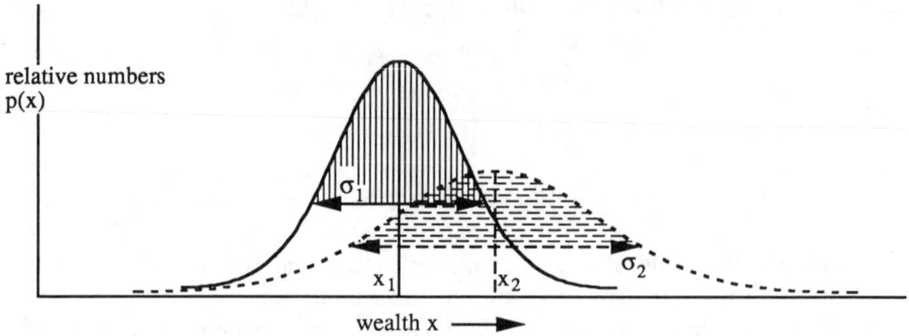

Figure 1.11 The relative number of people, $p(x)$, with wealth x given by the bell-shaped distribution for a narrow spread of standard deviation (σ_1) and broader spread (σ_2).

ber, $\int p(x)\, dx = 1$, constraint of average wealth x_0, $\int xp(x)\, dx = x_0$, and specified variance of wealth about the average, $\int (x - x_0)^2\, p(x)\, dx = \sigma^2$. σ (sigma) is the standard deviation or spread of the curve around its average value x_0 (which is x_1 or x_2). The "wealth" entropy is

$$h = - \int_0^\infty p(x) \log[p(x)]\, dx$$

$$= - \frac{1}{\sqrt{2\pi}\,\sigma} \int e^{-(x-x_0)^2/2\sigma^2} \left[\log\left(\frac{1}{\sqrt{2\pi}\sigma}\right) - \frac{(x - x_0)^2}{2\sigma^2}\right] dx$$

which evaluates (including the small cut-off tail for $x < 0$) to

$$h = \log \sigma + (\log \sqrt{2\pi} + \tfrac{1}{2}) \tag{1.62}$$

Curve 2 has the greater entropy—the more probable distribution of wealth—because σ_2 around $x_2 > \sigma_1$ around x_1 in Fig. 1.11.

Does this result mean the middle class is shrinking? This depends on how one defines "middle class." If it is the population within a specified fraction of the standard deviation it remains fixed, provided the bell-shaped curve is accurate. If the middle class is defined by the area within ± 25 percent, say, of the mean value, then it is tending to shrink with time. This *assumes* the mean or average wealth, after adjustment for inflation, rises slowly with time compared to the rate at which the standard deviation spreads.

These statistical statements do not preclude individuals from moving up the poor side of the solid curve to the rich side of the dashed curve later, or vice versa. The principle of increasing entropy merely implies a tendency for the poor segment of the population to grow poorer relative to the average wealth, and the rich segment to grow relatively richer as time progresses.

REFERENCES

[1.1] J.G. Ables, "Maximum entropy spectral analysis," In *Modern Spectrum Analysis*, D.G. Childers (ed). IEEE Press, 1978, pp. 23–31.

[1.2] C. Lanczos, *Linear Differential Operators*, New York: Van Nostrand, 1961, p. 132.

[1.3] R.M. Bevensee, "Fundamental limitations in antenna resolution by maximum entropy methods," Open Symposium on Data, Signal, and Image Processing, URSI XXI General Assembly, Florence, Italy, Aug. 28–Sept. 5, 1984.

[1.4] F.W. Sears, *Thermodynamics*, Cambridge, MA: Addison-Wesley, Chaps. 14–16, 1950.

[1.5] R.M. Bevensee, "Uncertainties in maximum entropy (ME) reconstructions," In *Digital Signal Processing-87*, V. Cappellini and A.G. Constantinides (eds.). Elsevier Science Publishers B.V. (North Holland), 1987, pp. 458–460.

[1.6] M. Kendall and A. Stuart, *The Advanced Theory of Statistics*, Vol. 1, 4th ed., New York: Macmillan, 1977.

[1.7] A.M. Mood, *Introduction to the Theory of Statistics*, New York: McGraw Hill, 1950, p. 153.

[1.8] B.R. Frieden, "Dice, entropy, and likelihood," Proc. IEEE, *73*, Dec., 1985, pp. 1764–1770.

[1.9] J. Capon, "High-resolution frequency-wavenumber spectrum analysis," Proc. IEEE, *57*, pp. 1408–1418, Aug., 1969. Reprinted in *Modern Spectrum Analysis*, D.G. Childers, (ed). IEEE Press, 1978, pp. 119–129.

[1.10] R.T. Lacoss, "Data adaptive spectral analysis methods," *Geophys.*, *36*, pp. 661–675, Aug. 1971. Reprinted in *Modern Spectrum Analysis*, D.G. Childers (ed). IEEE Press, 1978, pp. 134–148.

[1.11] J.P. Burg, "The relationship between maximum entropy spectra and maximum likelihood spectra," *Geophys.*, *37*, pp. 375–376, Apr., 1972. Reprinted in *Modern Spectrum Analysis*, D.G. Childers (ed.), IEEE Press, 1978, pp. 132–133.

[1.12] J.E. Shore and R.W. Johnson, "Axiomatic derivation of the principle of maximum entropy and the principle of minimum cross-entropy," IEEE Trans. Info. Theory, *IT-26*, Jan., 1980, pp. 26–37.

[1.13] J.E. Shore and R.W. Johnson, "Properties of cross-entropy minimization," IEEE Trans. Info. Theory, *IT-27*, July, 1981, pp. 472–482.

[1.14] E.T. Jaynes, "Prior probabilities," IEEE Trans. Syst. Science and Cyber., *SSC-4*, Sept., 1968, pp. 227–241.

[1.15] E.T. Jaynes, "Information theory and statistical mechanics," Part I, *Phys. Rev.*, *106*, March, 1957, pp. 620–630; Part II, ibid., *108*, October, 1957, pp. 171–191.

[1.16] J.W. Gibbs, *Elementary Principles of Statistical Mechanics Developed With Special Reference to the Rational Foundation of Thermodynamics*, New York: Dover, 1960, (unabridged, unaltered replication of the work first published by the Yale University Press in 1902.)

[1.17] L. Brillouin, *Science and Information Theory*, 2nd ed., New York: Academic Press, 1962.

[1.18] E.O. Thorp, *Beat the Dealer*, New York: Vintage Books, 1966.

[1.19] J.L. Kelly, Jr., "A new interpretation of information rate," The Bell System Tech. Journal, 35, July, 1956, pp. 917–926.

[1.20] C.E. Shannon, "A mathematical theory of communication," Bell System Tech. Journal, 27, Oct., 1948, pp. 379–423, 623–656.

2

Maximum Entropy Estimates of The Density of The Earth

2.1 INTRODUCTION

This chapter illustrates the two basic versions of the ME method—the Gibbs and the Boltzmann—by estimating the density distribution $\rho(r)$ for the earth solely from three data: its mass M, radius R, and moment of inertia J. Obviously, this problem is vastly underdetermined. Both ME methods predict a density of $\rho = 3.25 \pm 0.3$ grams/cm^3 for the mantle at the earth's surface. At the earth's center, the Gibbs estimate is 13.6 grams/cm^3 for a subdivision of the earth into 100 equivolume shells, while the corresponding Boltzmann estimate is 11.2 grams/cm^3. These numbers bound an independent estimate of 13.0 grams/cm^3.

How the ME analysis can be extended for lateral variations of density is also shown, using more gravity coefficients.

In the Gibbs's method, a probability distribution $p(\bar{\rho})$ for $(\rho_1, \ldots, \rho_N)$ in the N equivolume shells which maximizes the Shannon entropy is obtained, whereupon the ME estimate of ρ at each r is the expectation $\langle \rho(r) \rangle$ over this probability distribution. This Gibbs's $p^{\text{ME}}(\rho_1, \ldots, \rho_N)\, d\rho_1, \ldots, d\rho_N$ does *not* predict the relative number of possible earths which have densities in the range $\rho_1 \pm d\rho_1/2, \ldots, \rho_N \pm d\rho_N/2$ because not all of the possible earths described by p^{ME} satisfy the constraints.

The Boltzmann method is formally based on Bayes's Theorem Eq. (1.36) if there is prior information about each macrostate of the object or Eq. (1.37) if there are no priors. The Boltzmann ME object is the one which can be realized the greatest number of ways, and this is proportional to

$P(O)P_O(I_m)$ in Eq. (1.36) or $P_O(I_m)$ in Eq. (1.37), where $O = O^{ME}$. The derivation of the Fermi-Dirac distribution function in Sec. 1.3.2 exemplifies the Boltzmann method using prior information (the Uncertainty Principle).

Here the Boltzmann method is used without priors to build the earth density $\rho(r)$ in each shell with a number of building blocks such that the arrangement of numbers of blocks in all the shells can be realized the greatest number of ways (maximum H).

To obtain a probability distribution for possible $\rho(r)dV$ around $\rho^{ME}(r)dV$ in the Boltzmann method, only arrangements of building blocks must be counted which satisfy the constraints.

2.2 THE GIBBS METHOD—SPHERICALLY SYMMETRIC EARTH

The essence of the Gibbs ME method of data inversion is the assumption of an underlying joint probability distribution $p(\bar{\rho})$ for an ensemble of possible earths. The entropy and constraints are formulated in terms of this $p(\bar{\rho})$, which is discretized to $p(\rho_1, \ldots, \rho_N)$ for the N equivolume shells into which the earth is divided for the purpose of simplifying the ME inversion problem. N is a parameter which is varied, and the ME density becomes quite independent of it as N increases above 100. The following development is that of E. Rietsch [2.1].

The mass M is discretized for these N shells as

$$M = 4\pi \int_0^R \rho(r) r^2 \, dr = \frac{4\pi}{3} \sum_{n=1}^{N} \rho_n (r_n^3 - r_{n-1}^3) = \Delta V \sum_n \rho_n \tag{2.1}$$

with $\Delta V = 4\pi R^3/3N$, in terms of the true (unknown) shell densities ρ_n. The moment of inertia about a diameter is

$$J = \frac{8\pi}{3} \int_0^R \rho(r) r^4 \, dr = \frac{8\pi}{15} \sum_{n=1}^{N} \rho_n (r_n^5 - r_{n-1}^5) \tag{2.2}$$

with

$$r_n = R \left(\frac{n}{N}\right)^{1/3} \tag{2.3}$$

as the outer radius of the nth equivolume shell.

The Gibbs or Shannon entropy[1] is

$$\begin{aligned} h &= -\int d\bar{\rho}\, p(\bar{\rho}) \log[p(\bar{\rho})] \\ &= -\int d\rho_1 \cdots d\rho_N\, p(\rho_1, \ldots, \rho_N) \log[p(\rho_1, \ldots, \rho_N)] \end{aligned} \tag{2.4}$$

[1] Rietsch started with $\log[p(\bar{\rho})/w(\bar{\rho})]$ in Eq. (2.4), where $w(\bar{\rho})$ is a product of the $w_n(\bar{\rho}_n)$ which corresponds to the $r(x)$ as an invariant measure in Eq. (1.48). However, he set $w = 1$ for this problem.

Note these integrals are over densities of the shells, not spatial volume. h is to be maximized subject to the constraint

$$\int d\rho_1 \cdots d\rho_N p(\rho_1, \ldots, \rho_N) = 1 \tag{2.5}$$

The expectation values of the ρ_n, denoted $\langle\rho_n\rangle$ (in place of Rietsch's $\overline{\rho}_n$)

$$\langle\rho_n\rangle = \int d\rho_1 \cdots d\rho_N \rho_n p(\rho_1, \ldots, \rho_N) \tag{2.6}$$

should be consistent with the measured average earth density $\rho_{av} = M/V$, V being the earth volume. From Eq. (2.1), with ρ_n set equal to $\langle\rho_n\rangle$, this constrains

$$\sum_n \langle\rho_n\rangle = \frac{M}{\Delta V} = \frac{M}{V} N = \rho_{av} N \tag{2.7}$$

and from Eqs. (2.2) and (2.3), with $\rho_n = \langle\rho_n\rangle$,

$$\sum_n \langle\rho_n\rangle \, [n^{5/3} - (n-1)^{5/3}] = y\rho_{av} N^{5/3} \tag{2.8}$$

y has been conveniently defined as

$$y = \frac{5J}{2MR^2} = 0.84 \tag{2.9}$$

Maximization of h in Eq. (2.4) subject to Eqs. (2.5) to (2.8) is carried out in Appendix A, analogous to the derivations of the three distribution functions in Sec. 1.3. The result is

$$p(\overline{\rho}) = p(\rho_1, \ldots, \rho_N) = \prod_{n=1}^{N} p_n(\rho_n) \tag{2.10}$$

which implies independence of ρ_i and ρ_j in any two cells i and j,

$$p_n(\rho_n) = \frac{\exp[-\rho_n \eta_n(\overline{\beta})]}{Z_n(\overline{\beta})} \tag{2.11}$$

where

$$\eta_n(\overline{\beta}) = \beta_1 + \beta_2 \frac{[n^{5/3} - (n-1)^{5/3}]}{N^{2/3}} \tag{2.12}$$

and

$$Z_n(\overline{\beta}) = \int d\rho_n \exp[-\rho_n \eta_n(\overline{\beta})] \tag{2.13}$$

As discussed in Appendix A, each ρ_n can be restricted between a lower limit $\rho_l > 0$ and an upper limit $\rho_u < N\rho_{av}$. Then, ρ_n can be evaluated from

Eq. (2.6) as the expectation

$$\langle \rho_n \rangle = \int_{\rho_l}^{\rho_u} d\rho_n \rho_n p_n(\rho_n) \tag{2.14}$$

in terms of ρ_l, ρ_u, β_1, and β_2. The constraint Eqs. (2.7) and (2.8) yield β_1 and β_2. Then, the ρ_n are completely specified; they are the ME estimates of the densities of the N spherical shells.

A few comments are in order. The indeterminant nature of the earth is represented by a probability distribution of the shell densities $\rho_1, \ldots, \rho_N$. The Gibbs entropy as an integral of a "point" probability function $p(\overline{\rho})$ takes the following form for this problem

$$\begin{aligned} h &= -\int d\rho_1 \cdots \int d\rho_N \left[\prod_n p_n(\rho_n)\right]\left[\sum_m \log\left[p_m(\rho_m)\right]\right] \\ &= -\sum_m \int d\rho_1 \cdots \int d\rho_N \left[\prod_n p_n(\rho_n)\right] \log[p_m(\rho_m)] \\ &= -\sum_m \int d\rho_m p_m(\rho_m) \log[p_m(\rho_m)] \end{aligned} \tag{2.15}$$

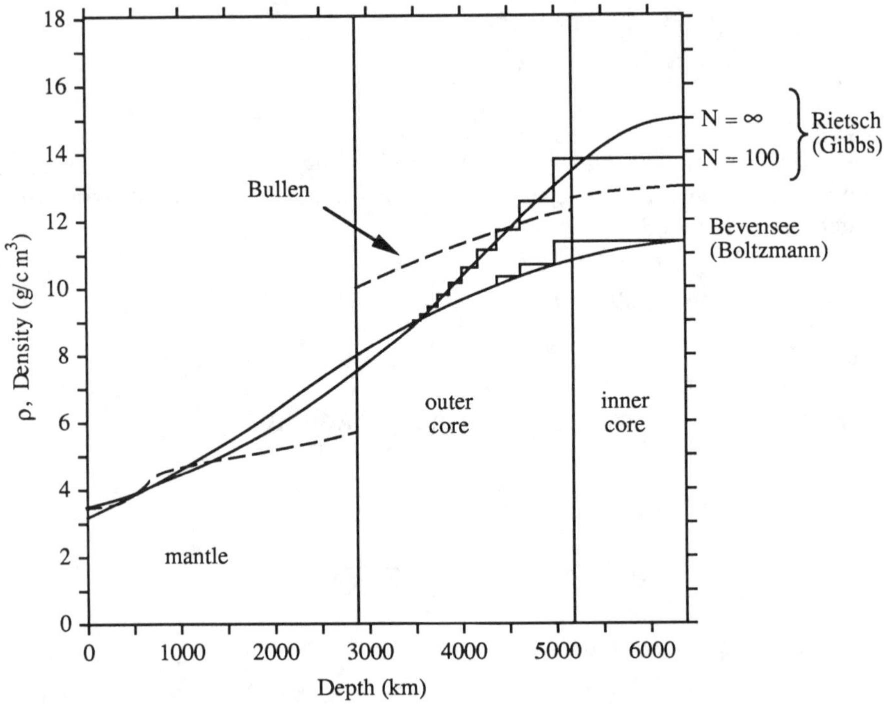

Figure 2.1 Density of the earth, $\rho(r)$, versus depth (R-r) as computed by ME procedures. R is the earth radius and r a distance from the center. Bullen's 1975 density distribution is shown for comparison. N = 100 denotes 100 equivolume shells. From Ref. [2.1] Fig. 1.

where $p_m = \rho_m/\sum \rho_n$ measures the fraction of total density in the mth shell (i.e. the fraction of total mass in the mth equivolume shell). The Gibbs constraints are in terms of the ME ρ_m averaged over their joint probability distribution, whereas the Boltzmann constraints will be in terms of the ρ_m alone.

Philosophically, there is no reason to prefer either the Gibbs or the Boltzmann version of ME for this earth-density problem, but synthetic data examples might favor a choice.

Rietsch's $\langle \rho_n \rangle$ density distribution obtained for a lower limit of $\rho_l = 0$ and no upper limit is shown in Fig. 2.1, both for $N = 100$ and $N = \infty$. The Boltzmann ME density to be derived in Sec. 2.4 is shown for comparison. It is interesting that they bound a 1975 estimate derived by K. Bullen [2.2].

2.3 EXTENSIONS FOR LATERALLY VARYING EARTH

The preceding formalism may easily be extended to include additional constraints of the form

$$F_q = \int_V \rho(\bar{r})\, f_q(\bar{r})\, dV, \qquad q = 1, \ldots, Q \qquad dV = dxdydz \tag{2.16}$$

where F_q is a gravity field coefficient and $f_q(\bar{r})$ is of the form $r^l Y_{lmi\cdot}(\theta,\phi)$ satisfying Laplace's equation. Y_{lmi} is a spherical harmonic of degree l, azimuthal order m and type $i = 1$ or 2 [cos(mϕ) or sin(mϕ)]. When Eq. (2.16) is discretized for N cells of volume dV $\rho(\bar{r})dV$ is defined as ρ_n and we have

$$F_q = \sum_n \rho_n f_q(r_n) \tag{2.17}$$

It has the form of Eq. (2.1) or (2.2). A Gibbs probability distribution $p(\bar{\rho}) = p(\rho_1, \ldots, \rho_N)$ is again postulated and the observed data $\langle F_q \rangle$ are presumed to be related to this distribution as

$$\begin{aligned} \langle F_q \rangle &= \int d\rho_1 \cdots \int d\rho_N p(\rho_1, \ldots, \rho_N) F_q \\ &= \sum_n \int d\rho_1 \cdots \int d\rho_N \rho_n p(\rho_1, \ldots, \rho_N) f_q(r_n) \end{aligned} \tag{2.18}$$

The objective function F is written analogous to Eq. (A.1), abbreviating $\rho_1, \ldots, \rho_N$ as $\bar{\rho}$,

$$\begin{aligned} F = -\int d\bar{\rho} p(\bar{\rho}) \log[p(\bar{\rho})] &+ \beta_0 \left[\int d\bar{\rho} p(\bar{\rho}) - 1\right] \\ &+ \sum_{q=1}^{Q} \beta_q \left[\sum_n \int d\bar{\rho} \rho_n p(\bar{\rho}) f_q(r_n) - \langle F_q \rangle\right] \end{aligned} \tag{2.19}$$

We set a variation $\delta F = 0$ for each of the independent $\delta p(\bar{\rho})$ and obtain the functional form of $p(\bar{\rho})$ by the procedure in Appendix B as

$$p(\bar{\rho}) = A_0 \prod_n e^{-R_n \rho_n}, \qquad R_n = \sum_{q=1}^{Q} \beta_q f_q(r_n) \ \text{mass}^{-1} \tag{2.20}$$

All the β_q, $1 \le q \le Q$, are determined by the set of constraints

$$\sum_{n=1}^{N} \frac{f_q(r_n)}{R_n} = \langle F_q \rangle, \qquad q = 1, \ldots, Q \tag{2.21}$$

A_0 is then found to normalize $p(\bar{\rho})$,

$$A_0 \prod_n R_n^{-1} = 1, \qquad A_0 \sim \text{mass}^{-N} \tag{2.22}$$

Both Eqs. (2.21) and (2.22) have been derived from Eq. (2.20) assuming no upper or lower limits to ρ for each cell, but this is not necessary.

The corresponding Gibbs ME estimate of each ρ_n (i.e., mass in each ΔV cell) is

$$\langle \rho_n \rangle = \int_0^{\infty} \rho_n p(\bar{\rho}) d\bar{\rho} = R_n^{-1} \tag{2.23}$$

and these can be computed from the $f_q(r_n)$ of the $n = 1, \ldots, N$ cells into which the earth is divided once the f_q are computed from the Q constraint equations Eq. (2.21). Note that one need not compute A_0 for Eq. (2.23).

An analysis similar to the foregoing was carried out with GEM spherical harmonic gravity field coefficients up to degree and order 36 as constraints [2.3]. The spherically symmetric portion of the earth density was assumed known and the lateral density distribution small compared to it. The resultant density distribution was computed for several crustal earth models, and all indicated broad density anomalies extending deep into the mantle, which is about 3000 km thick.

This particular analysis is interesting because it is a "hybrid" type, neither pure Gibbsian like the previous one nor pure Boltzmann like the one in the next section. Entropy h was formulated in terms of the probability distribution P_i of i earth models instead of $p(\rho_n)$ indexed by the densities of the earth cells. Objective function F was made stationary, thereby maximizing the entropy, by variations of the P_i rather than the $p(\rho_n)$. A necessary assumption in the procedure seemed to be no limits on the number of particles (i.e., density) occupying each earth cube, which limits are allowed in a pure Gibbsian analysis though not easily included in the Boltzmann procedure. The final result was a density distribution in position $\bar{r}$ which had the form of an unnormalized Boltzmann distribution but which had been derived by a Gibbsian averaging over various earth models with their P_i probabilities.

We feel that the conclusions from such a hybrid analysis may be less realistic than from either a pure Gibbsian or pure Boltzmann analysis, although further study is required to resolve this issue.

2.4 THE BOLTZMANN METHOD—SPHERICALLY SYMMETRIC EARTH

The mass M and moment of inertia J are discretized according to Eqs. (2.1) to (2.3). To set up the Boltzmann entropy we regard each density ρ_n as constructed of an integral number of tiny building blocks $\hat{d}\rho$

$$\rho_n = M_n \hat{d}\rho \tag{2.24}$$

The number of distinct ways of forming the arrangement (macrostate) of $M_1, \ldots, M_N$ blocks is the number of combinations C from Eq. (1.3)

$$C = \frac{M_T!}{\prod_n M_n!}, \qquad M_T = \sum_n M_n \tag{2.25}$$

Entropy H is log(C), given by Eq. (1.8) for large numbers M_n, as

$$H = \log C = -\sum_n M_n \log\left(\frac{M_n}{M_T}\right) \tag{2.26}$$

It is convenient to form the objective function using the densities ρ_n, so we substitute $M_n = \rho_n/\hat{d}\rho$ and ignore the overall factor of $1/\hat{d}\rho$ in H. Alternatively, $\hat{d}\rho$ could be absorbed into the Lagrange multipliers. Then, H becomes

$$H = \hat{d}\rho \log C = -\sum_n \rho_n \log\left(\frac{\rho_n}{\rho_T}\right) \tag{2.27}$$

Constraint Eq. (2.1) with Eq. (2.3) can be rewritten as

$$\rho_T = \sum \rho_n = \rho_{av} N \tag{2.28}$$

which corresponds formally to the Gibbs expression Eq. (2.7), but now ρ_n is a direct estimate of the density in the nth shell instead of $\langle \rho_n \rangle$ as an average over a Gibbs probability distribution. Constraint Eq. (2.2) is rewritten with Eq. (2.3) as

$$\sum_n \rho_n[n^{5/3} - (n-1)^{5/3}] = y\rho_{av} N^{5/3} \tag{2.29}$$

which formally corresponds to the Gibbsian Eq. (2.8) except that ρ_n replaces the Gibbs $\langle \rho_n \rangle$.

We form the Boltzmann objective function F by writing H of Eq. (2.27) plus the constraint from Eq. (2.28) multiplied by a Lagrange multiplier β_1, plus the constraint from Eq. (2.29) times β_2. Then,

$$F = -\sum \rho_n \log\left(\frac{\rho_n}{\rho_T}\right) + \beta_1 [\sum_n \rho_n - \rho_{av}N] + \beta_2 \left\{\sum_n \rho_n[n^{5/3} - (n-1)^{5/3}] - y\rho_{av}N^{5/3}\right\} \tag{2.30}$$

Compare this with the Gibbs objective function (A.1) of Appendix A. The Gibbs approach maximizes the entropy of the underlying probability distribution for the probability $p(\rho_1, \ldots, \rho_N)$, whereas the Boltzmann method maximizes the entropy of the spatial $\bar{\rho}$ distribution directly. Also, the Gibbs constraints are in terms of the expected values of the ρ_n, $\langle\rho_n\rangle$, averaged over $p(\bar{\rho})$, whereas the Boltzmann constraints are expressed directly in terms of the spatial $\bar{\rho}$ distribution.

When we set $\delta F = 0$ with respect to every $\delta\rho_i$ in Eq. (2.30) we obtain easily (the total ρ, ρ_T, is fixed)

$$\rho_n = \rho_T e^{-1+\beta_1} \exp\left\{\beta_2[n^{5/3} - (n-1)^{5/3}]\right\} \tag{2.31}$$

We can evaluate $e^{-1+\beta_1}$ very easily from constraint (2.28) by summing (2.31) over n. Then

$$\rho_n = \rho_T \frac{\exp\{\beta_2[n^{5/3} - (n-1)^{5/3}]\}}{Z(\beta_2)} \tag{2.32}$$

with

$$Z(\beta_2) = \sum_n \exp\{\beta_2[n^{5/3} - (n-1)^{5/3}]\} \tag{2.33}$$

Compare this with the product of the Gibbs $Z_n(\bar{\beta})$-factors in Eq. (2.13). It is very difficult to include upper or lower bounds on the ρ_n in the Boltzmann method, whereas it is very straightforward in the Gibbs procedure.

To evaluate β_2 we insert Eqs. (2.32) and (2.33) into the second constraint Eq. (2.29). With $y = 0.84$ of Eq. (2.9) we obtain the condition

$$\sum_n \left\{\left(\frac{n}{N}\right)^{5/3} - \left[\frac{(n-1)}{N}\right]^{5/3}\right\} \exp\{\beta_2[n^{5/3} - (n-1)^{5/3}]\} = \frac{0.84}{N} \sum_n \exp\{\beta_2[n^{5/3} - (n-1)^{5/3}]\} \tag{2.34}$$

For $N = 100$, trial-and-error solution of this for β_2 gives us the Boltzmann ME density of the nth shell as

$$\rho_n^{ME} = 11.63 \exp\left\{-80.97\left[\left(\frac{n}{100}\right)^{5/3} - \left(\frac{n-1}{100}\right)^{5/3}\right]\right\}, \qquad 1 \le n \le 100 \tag{2.35}$$

This is plotted on Fig. 2.1 and underbounds the density distribution estimated by Bullen at the lower depths, whereas the Gibbs ME density distribution $\langle\rho_n\rangle$ of Eq. (2.14) overbounds the Bullen estimate.

2.5 INTERPRETATION OF THE ME PROBABILITY DISTRIBUTIONS FOR DENSITY $\rho(r)$ RELATIVE TO THE TRUE DENSITY

We continue to assume a spherically symmetric earth divided into N equivolume shells for these considerations.

2.5.1 The Gibbs's $p(\bar{\rho}) = \prod_n p_n(\rho_n)$, with Density ρ_n at r_n

In an ensemble of *possible* earths $p(\bar{\rho})d\bar{\rho}$, gives the fractional number of them within the intervals $\rho_1 \pm d\rho_1/2, \ldots, \rho_N \pm d\rho_N/2$ without regard for the constraints of mass M and moment of inertia J. The ME density distribution $\langle\bar{\rho}\rangle^{\text{ME}} = \langle\rho_1, \ldots, \rho_N\rangle^{\text{ME}}$, averaged over $p^{\text{ME}}(\bar{\rho})$ was derived so as to satisfy these constraints. The question now arises: Can we use this Gibbs's ME $p(\bar{\rho})$ directly to assure ourselves in some sense that a true value ρ_i^{true} in the ith shell lies within a specified interval $\Delta\rho_i$ of $\langle\rho_i\rangle^{\text{ME}}$?

Since the Gibbs's probability distribution $p(\bar{\rho})$ includes *all possible* $\bar{\rho}$ the equivalent question is: How many of the $\bar{\rho}$ ensemble have ρ_is which lie within this interval $\Delta\rho_i$?

In Sec. 3.8, Assurance is defined as the probability that a member of an ensemble of possible earths *all obeying the constraints* has a ρ_i which falls within the interval $\Delta\rho_i$. One might consider computing this Assurance by an integral of the Gibbs's $p(\bar{\rho})$ over this interval of ρ_i as

$$\int \prod_{n\neq i} d\rho_n \int_{\langle\rho_i\rangle^{\text{ME}}-\Delta\rho_i}^{\langle\rho_i\rangle^{\text{ME}}+\Delta\rho_i} p^{\text{ME}}(\bar{\rho})\, d\rho_i = \int_{\langle\rho_i\rangle^{\text{ME}}-\Delta\rho_i}^{\langle\rho_i\rangle^{\text{ME}}+\Delta\rho_i} p_i^{\text{ME}}(\rho_i)\, d\rho_i \quad (2.36)$$

But this would be incorrect because a typical subensemble of the population, $(\rho_1' \pm d\rho_1/2, \ldots, \rho_N' \pm d\rho_N/2)$, *does not satisfy the constraints*. Only the expected value $\langle\rho_1, \ldots, \rho_N\rangle^{\text{ME}}$ satisfies the constraints.

For this reason the Gibbs's ME probability distribution must be employed with care to obtain Assurance(s) that ME-estimated parameters fall within specified range(s) and to obtain distribution functions for these parameters over the subensemble of systems which satisfy the constraints.

2.5.2 The Boltzmann $p(\bar{\rho}) = \prod_k (M_k/M_T)$, where $\rho_k = M_k\hat{d}\rho$ and $M_T = \sum M_k$

As was the case in Sec. 2.5.1, we must compute an Assurance that the true density ρ_i^{true} lies within a specified interval of $\rho_i^{\text{ME}} = M_i^{\text{ME}}\,\hat{d}\rho$ by counting in the ensemble of possible densities ($\rho_1, \ldots, \rho_N$) only those within the interval which satisfy the constraints. In Sec. 3.8 this is accomplished with the aid of the Darwin-Fowler theory. This will yield a Poisson-like probability distribution for ρ_i at r_i about ρ_i^{ME}. All possible earths included in this probability distribution satisfy the constraints.

The Assurance that ρ_i^{true} lies within a specified interval of ρ_i^{ME} is precisely the integral of this Poisson-like probability distribution over the interval. The value depends on the "discrimination level" or size of the building block, $\hat{d}\rho$, used to construct the Boltzmann density distribution $\bar{\rho}$. The smaller $\hat{d}\rho$ is, the tighter will be the Poisson-like distribution around ρ_i^{ME} and the higher the Assurance will be that ρ_i^{true} lies close to ρ_i^{ME}. Larger $\hat{d}\rho$ implies poorer resolution and smaller Assurance that ρ_i^{true} lies within a specified interval around ρ_i^{ME}.

These matters are elaborated on in Sec. 3.8 and 3.9.

REFERENCES

[2.1] E. Rietsch, "The maximum entropy approach to inverse problems, *J. Geophys.* (Springer-Verlag, Heidelberg) *42*, 1977, pp. 489–506.

[2.2] K. E. Bullen, *The Earth's Density*, London: Chapman and Hall, 1975.

[2.3] D.P. Rubincam, Information theory lateral density distribution for earth inferred from global gravity field, *Jour. Geophysical Research*, Vol. *87*, No. B7, July 10, 1982, pp. 5541–5552.

3

Maximum Entropy Geotomography (Boltzmann Method)

3.1 INTRODUCTION

Tomography is the practice of constructing a three-dimensional parameter distribution within an underdetermined body from ray data. A practical example is the reconstruction of the density distribution of tissue and bone within part of a human body by "slices." Through one slice at a time enough rays (usually microwaves or X rays) are transmitted in different directions to enable a density reconstruction with adequate spatial resolution. In this chapter we tackle the problem of reconstructing a volumetric parameter distribution of density, electrical conductivity, or perhaps dielectric constant within the earth by ray data. The problem is made two dimensional by considering just a vertical rectangular slice of earth discretized into small cells. On each vertical boundary of this slice we assume two holes have been bored. In one of the holes a transmitter is placed at various depths. At each position it generates an acoustic or electromagnetic wave which probes the region on its way to the other borehole, where receivers at different depths record the times of arrival, amplitudes, or perhaps phases of the arriving rays. Since the number of cells to be resolved is invariably much greater than a practical number of rays, the inversion problem is underdetermined, and is a natural candidate for the Boltzmann ME technique.

This problem is simpler than the spectral superresolution problem of Chap. 5 because our earth data are real rather than complex.

In Sec. 3.2, the problem is formulated as a discretized version of the Fredholm integral equation of the first kind. This integral equation is well known to be ill-conditioned in that small changes in the ray data (image) can cause large changes in the unknown parameter distribution (object) under the integral sign. The discretized Fredholm equation furnishes one constraint for each ray datum in terms of the parameter amplitudes in the cells traversed by that ray. A given parameter distribution throughout the cells is assumed to be built with an enormous number of tiny building blocks, and the number of blocks n_1 in the first cell, . . . , n_k in the kth cell, . . . is found which maximizes the entropy, log C, C being the number of distinct combinations of these numbers, subject to the ray constraints and $\sum n_k = N$ (total number of building blocks). This procedure is entirely analogous to the derivation of the discrete density distribution in the equivolume shells comprising the earth model in Sec. 2.4.

In Sec. 3.3 an algorithm is described for iterating to the ME parameter distribution within the cells, starting from an initial guess of the Lagrange multipliers and the constraint $x_T = \sum x_k$, x representing the parameter of interest (density, electrical conductivity, or dielectric constant). This algorithm has proved to be generally successful for synthetic-data as well as real-data problems, provided the number of rays is $\lesssim$ half the number of cells. This means the algorithm is useful for grossly undetermined problems.

In Sec. 3.4 some of the instability problems are described which can arise in the numerical inversion procedure with a poor choice of rays or too many of them. These problems also arise when one wants to estimate the changes in an x_k distribution from small changes in the ray data (that is, effect of noise). We derive a criterion for a well-behaved algorithm to converge on the ME x_k distribution and allow computation of the perturbing effect of noise.

In Sec. 3.5 a relaxation technique based on eigenvalue decomposition is described, which has proven useful in synthetic-data examples when the process of updating the Lagrange multipliers to improve the ME estimates of the ray data is unstable. This situation often arises when the number of ray data is larger than necessary for convergence to one parameter distribution.

In Sec. 3.6, the ME solutions to several synthetic-data examples are described to give the reader a "feel" for the accuracy of the method.

In a practical geotomographic problem, the rays bend or diffract and we comment on how one might modify the algorithm of Sec. 3.3 to correct the ray paths each iteration of the x- and β-parameters toward their ME values.

A ME object is only a most-probable estimate of the true object obtained from its image, and so we make, in Sec. 3.8, some estimates of how far away the true object might be from that estimate. Even with no image noise and error-free data processing there is statistical uncertainty in the true-object estimate simply because many objects can realize the measured image. Jaynes's Concentration Theorem tells us the probability of the object lying within an entropy interval δh of h_{max}. A Darwin-Fowler analysis yields a probability distribution for an object parameter x_k around its x_k^{ME} value, provided the continuous x parameter is quantized into units of $\hat{d}x$. Assurance is defined as the probability that the true x_k lies within a specified interval of x_k^{ME}, independent of any noise that might be present.

In Sec. 3.9 the general, difficult problem of noiseless object estimation is treated from a noisy image. The noisy-image region is mapped by the ME inversion process into a noisy-object region, with the statistical uncertainty superimposed on each possible noisy-x_k^{ME} object. The noise variance of an x_k parameter about its reference value is computed from the autocorrelations of the image cell noises, and the x_k statistical uncertainty variance is added to obtain the total variance. The *Confidence* that the true x_k lies within an interval of its reference ME value is defined as the total probability of all possible values of x_k falling within that interval.

3.2 FORMULATION FOR TWO-DIMENSIONAL PROBLEMS

Figure 3.1 shows a two-dimensional vertical slice of the underground, discretized into a number K of small cells. A continuous distribution of parameter $x(\bar{r})$ is also discretized, x_k being the unknown value in the kth cell. To estimate this distribution, two vertical boreholes have been drilled and various rays have been launched from the indicated points along the left borehole, traveling to the right. The rays could be wavefronts of acoustic pressure waves to measure inverse wave velocity (which yields density), or radio frequency electromagnetic waves to measure electrical conductivity or dielectric constant. Inverse wave velocities in the cells would be inferred by times of arrival of the pressure waves; conductivity, by the path attenuations of the electromagnetic waves; and dielectric constant, by the phase shifts of the rays. The symbol T_k (a real number) represents the kth ray datum.

We assume straight rays until Sec. 3.7, where the problem of accounting for ray bending or diffraction is discussed.

The Fredholm equation of the first kind [3.1], appropriate for this problem, is

$$T(\bar{r}) = \int \mathfrak{D}(\bar{r}, \bar{r}')x(\bar{r}')dV' \tag{3.1}$$

where $T(\bar{r})$ is the observable effect (image) at point $\bar{r}$, $x(\bar{r}')$ is the unknown parameter distribution (object) at $\bar{r}'$ and $\mathfrak{D}$ is the known kernel. For example, the integrand could be $dx'/v(\bar{r}')$, $v(\bar{r}')^{-1}$ being the inverse wave velocity, for a ray passing through on its way to $\bar{r}$ with time of arrival $T(\bar{r})$. Formally, we discretize Eq. (3.1) by expanding x as

$$x(\bar{r}') = \sum_{k=1}^{K} x_K\, f_K(\bar{r}') \tag{3.2}$$

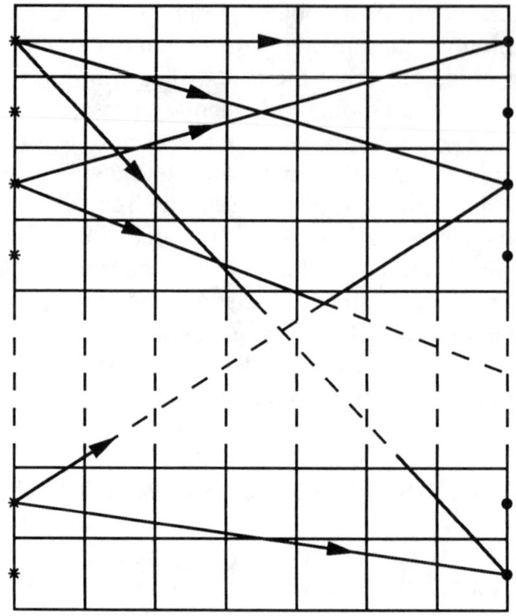

Figure 3.1 A two-dimensional vertical slice of the underground, discretized into K cells. There are no variations in the direction perpendicular to the paper. Some possible rays are shown from transmitters in the left borehole to receivers in the right borehole.

$f_k(\bar{r}')$ being a localized expansion function, and by sampling Eq. (3.1) with a "testing" function $g_i(\bar{r})$ of limited spatial extent. The testing is done by multiplying Eq. (3.1) with a given $g_i(\bar{r})$ and integrating over $\bar{r}$. This yields

$$T_i = \sum_{k=1}^{K} D_{ik}\, x_k, \qquad 1 \le i \le I < K \tag{3.3}$$

with

$$T_i = \int g_i(\bar{r}) T(\bar{r}) dV \tag{3.4}$$

$$D_{ik} = \int dV \int dV' \; g_i(\bar{r})\, \mathcal{D}(\bar{r}, \bar{r}')\, f_k(\bar{r}') \tag{3.5}$$

For our geotomography problem, we take $g_i(r) = \delta(\bar{r} - \bar{r}_i)$ at the point $\bar{r}_i$ along the ith ray, and $f_k(\bar{r})$ a uniform pulse of amplitude 1 over the kth cell only. T_i, the datum of ray i, D_{ik}, the distance the ith ray travels in cell k, and x_k, the k-cell parameter, all have their intuitive meanings shown in

Fig. 3.1. The matrix **D**, of which D_{ik} is the element in row i, column k, is the path matrix: the ith row displays the path lengths in all cells 1, . . . , k, . . . , K.

We neglect not only ray bending because of the cell interface boundary conditions in a cellular medium but also reflection of rays at interfaces, across which the parameter x of interest is discontinuous (in this model).

Our ME procedure of solving the problem for the "most probable" x_k distribution [3.2] parallels that of Sec. 2.4, where the earth density in the kth spherical shell is analogous to x_k here. Now there are many more ray constraints than earth density constraints of total mass and moment of inertia. The steps of deriving the entropy expression and the objective function F are summarized.

Each x_k is regarded as constructed of an enormous number n_k of building blocks of size $\hat{dx}$. The magnitude of $\hat{dx}$ will not affect the ME solution, but it will play the role later of a statistical parameter in the distribution of possible x_k values about x_k^{ME}. The total amount of unknown parameter is denoted

$$x_T = \sum x_k, \qquad \text{and} \qquad \text{N} = \sum n_k \tag{3.6}$$

is the total number of building blocks. There is no a priori preference for particular values or bounds on the numbers n_k so the total number of discrete combinations of N blocks with n_1 in the first cell, . . . , n_k in the kth cell, . . . , is

$$\text{C} = \frac{\text{N}!}{\prod_k n_k!} \tag{3.7}$$

Figure 1.3 in Sec. 1.3.1 illustrates all the combinations C of "macrostates" among two cells of $n_1 = 3$, $n_2 = 1$; the four $\hat{dx}$ blocks are labeled by letters a, b, c, d.

The entropy H = log C reduces to

$$\text{H} = \log \text{C} = -\sum_k n_k \log\left(\frac{n_k}{\text{N}}\right) \tag{3.8}$$

and the objective function F is the entropy plus each constraint equation multiplied by its Lagrange multiplier:

$$\text{F} = \text{H} + \gamma\left(\sum_k n_k - \text{N}\right) + \sum_{i=1}^{\text{I}} \beta_i'\left(\sum_{k=1}^{\text{K}} \text{D}_{ik}\, n_k\, \hat{dx} - \text{T}_i\right) \tag{3.9}$$

Setting $\delta\text{F} = 0$, as usual, with respect to each δn_k in turn yields K equations for all the n_k. $\hat{dx}$ can be absorbed into the Lagrange multipliers by redefining each $\beta_i = \beta_i'\hat{dx}$. The remaining I + 1 equations for the K + I + 1 unknown x, β_i, and γ are furnished by the I constraint Eqs. (3.3) and (3.6).

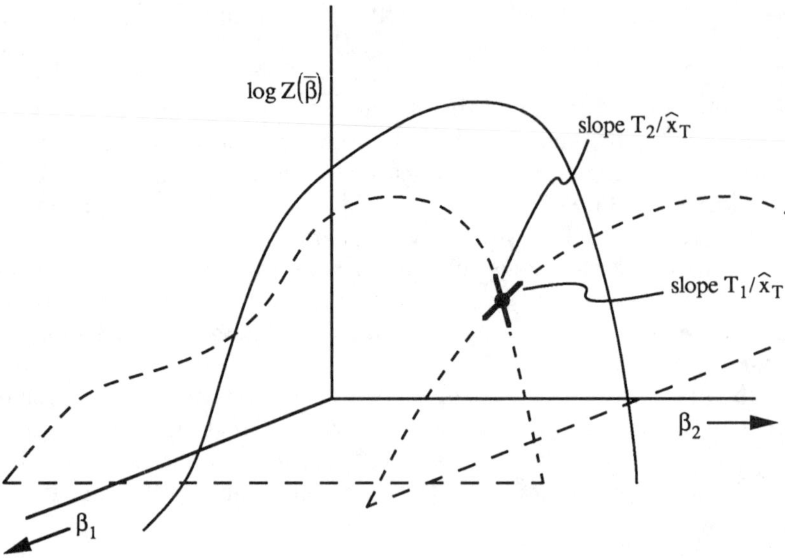

Figure 3.2 The solution of all the I equations Eq. (3.12) is represented by the intersection of all the β_i planes with the $\log Z(\bar{\beta})$ hypersurface. The β_i plane is perpendicular to the β_i axis. At the solution point (dot), all the constraints of Eq. (3.12) are satisfied.

The jth equation for $\hat{x}_j$, the current[1] ME estimate of x_j, obtained from this procedure is

$$\hat{x}_j = \hat{x}_T \exp\left[\frac{\sum_{i=1}^{I} \beta_i D_{ij}}{Z(\bar{\beta})}\right] \qquad j = 1, \ldots, K \tag{3.10}$$

where the "partition function" Z guarantees that Eq. (3.6) is obeyed,

$$Z(\bar{\beta}) = \sum_{k=1}^{K} \exp\left[\sum_i \beta_i D_{ik}\right] \tag{3.11}$$

The I constraint equations which implicitly determine all the $\beta_1, \ldots, \beta_I$ for the I paths are obtained by substituting Eq. (3.10) into Eq. (3.3)

$$\sum_{k=1}^{K} D_{jk} \frac{\exp\left[\sum_{i=1}^{I} \beta_i D_{ik}\right]}{Z(\bar{\beta})} = \frac{\hat{T}_j}{\hat{x}_T}, \qquad 1 \leq j \leq I < K \tag{3.12}$$

[1] The hat over a symbol denotes an estimate from the ME equations, which are iterated to a solution of acceptable accuracy in the constraints.

The left side is identically $\partial \log [Z(\bar{\beta})]/\partial\beta_j$. $\hat{T}_j$ is the current estimate of T_j.

This solution parallels that of Jaynes (3.3). The solution is maximum entropic because any set of variations $\delta n_j = \delta x_j/\hat{d}x$ about the n_j which maintains all the constraints does not change H, and $\partial^2 H/\partial n_k^2 < 0$ for all k.

The solution of the I equations Eq. (3.12) is shown by the dot in $\bar{\beta}$ hyperspace in Fig. 3.2.

The entropy H of Eq. (3.8) divided by N can be written as

$$\begin{aligned} h &= -\sum_k^{K} \left(\frac{\hat{x}_k}{\hat{x}_T}\right) \log \left(\frac{\hat{x}_k}{\hat{x}_T}\right) \\ &= \log Z(\bar{\beta}) - \sum_i^{I} \frac{\beta_i \hat{T}_i}{\hat{x}_T} \end{aligned} \tag{3.13}$$

in which $\hat{T}_i/x_T = \partial \log[Z(\bar{\beta})]/\partial\beta_i$.

If the initial estimate $\hat{x}_T^{(0)}$ of x_T is close to the correct value, the self-consistent solution to Eqs. (3.10) and (3.12) by the algorithm in the next section can yield the correct $\hat{x}_T = x_T$. This is so provided at least one ray passes through each of the K cells being resolved.

3.3 AN ALGORITHM FOR AN ITERATIVE SOLUTION [3.2]

We continue to neglect ray bending (diffraction) and reflection in the cells so the D_{ik} path lengths are known. Our algorithm starts with an initial guess $\beta_i^{(0)}$ of each β_i. Let us abbreviate all of these with the vector $\bar{\beta}^{(0)}$. We also start with a reasonable estimate of x_T, $\hat{x}_T^{(0)}$. The algorithm then proceeds as follows:

1. The estimates $\hat{x}_k^{(1)}$ are obtained from $\bar{\beta}^{(0)}$ by Eq. (3.10).
2. The estimates $T_j^{(1)}$ are computed from Eq. (3.12).
3. $\hat{x}_T^{(0)}$ is scaled to $\hat{x}_T^{(1)}$ by multiplying the former by

$$r = \frac{\left(\sum \beta_i^{(0)} T_i\right)}{\left(\sum \beta_i^{(0)} \hat{T}_i^{(1)}\right)}, \qquad \text{or} \qquad \frac{\sum T_i}{\sum \hat{T}_i^{(1)}} \quad \text{if all } \beta_i^{(0)} = 0 \tag{3.14}$$

4. All the $\hat{T}_i^{(1)}$ and $\hat{x}_j^{(1)}$ are also scaled by this factor

$$\hat{T}_i^{(1)} \to r\hat{T}_i^{(1)}, \qquad \hat{x}_k^{(1)} \to r\,\hat{x}_k^{(1)} \tag{3.15}$$

   This scaling does not change the values of entropy h in Eq. (3.13).

5. Next, we update $\bar{\beta}^{(0)}$ to bring the computed $\hat{T}_j$ of Eq. (3.12) closer to the measured T_j. We do this by solving for all the derivatives

$\partial T_i/\partial \beta_j = \partial T_j/\partial \beta_i$ by Eq. (3.12), then invert the set of I equations

$$\Delta T_i = T_i - \hat{T}_i = \sum_{j=1}^{I} \left(\frac{\partial \hat{T}_i}{\partial \beta_j}\right) \Delta \beta_j, \qquad 1 \leq i \leq I \tag{3.16}$$

for all the $\Delta \beta_j$, whereupon the improved $\overline{\beta}$ is $\overline{\beta}^{(1)} = \overline{\beta}^{(0)} + \overline{\Delta\beta}$.

We now have all the $\hat{x}_k^{(1)}$, $\hat{T}_j^{(1)}$, $\hat{x}_T^{(1)}$, and $\beta_j^{(1)}$ so we return to step (1) and compute $\hat{x}_k^{(2)}$ from $\overline{\beta}^{(1)}$, and so on, through as many iterations as necessary to obtain convergence of the $\hat{x}_k$, $\hat{x}_T$ and $\overline{\beta}$ with acceptable accuracy in the $\hat{T}_j$. The final answers should be independent of the starting $\overline{\beta}^{(0)}$, but may not quite be so because of limited computer word size.

As the iterations proceed, the matrix inversion of Eq. (3.16) to obtain the $\Delta \beta_j$ may become very ill-conditioned. The procedure of (LU) decomposition and back substitution in the matrix representing the $\partial \hat{T}_i/\partial \beta_j$ array may result in very small pivot elements on the diagonal. These can cause numerical overflow unless the computer word length is sufficiently large. For the problems we have encountered, the double-precision word length of 16 decimal digits on a VAX computer is adequate. Solutions of problems like those described in Sec. 3.6 to three significant digits or better occur typically in about ten iterations of the algorithm.

The numerical instability inherent in the inversion of the set Eq. (3.16) arises from the ill-conditioned nature of the Fredholm integral equation Eq. (3.1): a small change in the data, $T_i - \hat{T}_i$, can be caused by large changes in the "source" terms x_k, measured by the changes in the β_j. The instability becomes acute when the number of rays I is increased beyond a limit determined by the ME $\hat{x}_k$ distribution itself, as well as the path parameters D_{ik}. In the next section, the reasons for this behavior are examined and a criterion is developed to avoid it and facilitate a noise analysis as well.

3.4 STABILITY AND NOISE

The nature of Matrix $\mathbf{M}$ with components $\mathbf{M}_{ij} = \partial \hat{T}_i/\partial \beta_j$ in Eq. (3.16) becomes apparent by evaluating the derivative in Eq. (3.12) with $\hat{x}_T$ held fixed. $\hat{x}_T$ is the current best estimate of the total amount of parameter x_T. We obtain

$$\frac{\partial \hat{T}_i}{\partial \beta_j} = \sum_k D_{ik}\, x_k\, D_{jk} - \frac{1}{\hat{x}_T}\left(\sum_k D_{ik}\, \hat{x}_k\right)\left(\sum_{k'} D_{jk'}\, \hat{x}_{k'}\right) = \frac{\partial \hat{T}_j}{\partial \beta_i} \tag{3.17}$$

The first sum is

$$(\mathbf{DXD^t})_{ij}, \qquad \text{with } \mathbf{X} = \text{diag}(\hat{x}_1, \ldots, \hat{x}_K) \tag{3.18}$$

This is called an "intersection" matrix because the ijth term is the sum of all the products of the ith and jth path lengths in the cells, each product multiplied by the "x" in that cell. The second term can be written as

$$(\mathbf{Dxx^tD^t})_{ij}, \qquad \text{with } \mathbf{x} = \text{col}(\hat{x}_1, \ldots, \hat{x}_K) \tag{3.19}$$

So the matrix form of Eq. (3.16) is

$$\mathbf{\Delta T} = \left[\mathbf{D}(\mathbf{X} - \frac{1}{\hat{x}_T}\mathbf{xx^t})\ \mathbf{D^t} \right] \cdot \mathbf{\Delta \beta} \tag{3.20}$$

The eigenvalues λ_i of the K × K matrix **G** in parentheses

$$\mathbf{G} = \mathbf{X} - \frac{1}{\hat{x}_T}\mathbf{xx^t} \tag{3.21}$$

contain at least one zero value. For if we solve the eigenvalue equation

$$\mathbf{G\ W_k} = \lambda_k\ \mathbf{W}_k, \qquad (\mathbf{G} - \lambda\mathbf{U})\mathbf{W_k} = 0 \tag{3.22}$$

(**U** = unit matrix) we obtain a nontrivial solution for the set of λ_k and $\mathbf{W}_k$ = col($W_{k1}, \ldots, W_{kK}$) only if the determinant is

$$\det(\mathbf{G} - \lambda\mathbf{U}) = 0 \qquad \text{for each } \lambda_k \tag{3.23}$$

This can be written as

$$\det \begin{vmatrix} \hat{x}_1(1 - \hat{x}_1/\hat{x}_T) - \lambda & -\hat{x}_1\hat{x}_2/\hat{x}_T & -\hat{x}_1\hat{x}_3/\hat{x}_T & \cdots \\ -\hat{x}_2\hat{x}_1/\hat{x}_T & \hat{x}_2(1 - \hat{x}_2/\hat{x}_T) - \lambda & -\hat{x}_2\hat{x}_3/\hat{x}_T & \cdots \\ -\hat{x}_3\hat{x}_1/\hat{x}_T & -\hat{x}_3\hat{x}_2/\hat{x}_T & \cdot & \cdots \\ \vdots & \vdots & \ddots & \cdots \end{vmatrix} = 0 \tag{3.24}$$

If we add rows 2,3, . . . , K to row 1 (which doesn't change the λ_i eigenvalues) and note $\hat{x}_k\ (1 - \hat{x}_k/\hat{x}_T) - \sum_{k' \neq k} \hat{x}_{k'} \cdot \hat{x}_k/\hat{x}_T = 0$ in every column k we obtain

$$\det \begin{vmatrix} -\lambda & -\lambda & -\lambda & \cdots \\ -\hat{x}_2\hat{x}_1/\hat{x}_T & \hat{x}_2\ (1 - \hat{x}_2/\hat{x}_T) - \lambda & -\hat{x}_2\hat{x}_3/\hat{x}_T & \cdots \\ -\hat{x}_3\hat{x}_1/\hat{x}_T & -\hat{x}_3\hat{x}_2/\hat{x}_T & \hat{x}_3\ (1 - \hat{x}_3/\hat{x}_T) - \lambda & \cdots \\ \vdots & \vdots & \vdots & \ddots \end{vmatrix} = 0 \tag{3.25}$$

This is satisfied by $\lambda = 0$. Hence, **G** of Eq. (3.21) has at least one zero eigenvalue in matrix $\mathbf{\Lambda}$ = diag ($\lambda_1, \ldots, \lambda_K$).

We can write the columns $\mathbf{W}_1, \mathbf{W}_2, \ldots$ as one K × K matrix **W**, where

$$\mathbf{GW} = \mathbf{W\ \Lambda} \tag{3.26}$$

so the spectral representation of **G** is, with the normalization $\mathbf{W^t} \cdot \mathbf{W} = \mathbf{U}$,

$$\mathbf{G} = \mathbf{W\Lambda W^t}, \qquad \mathbf{W^t} = \mathbf{W^{-1}} \tag{3.27}$$

With this, Eq. (3.20) for a small change $\mathbf{\Delta T}$ due to a change $\mathbf{\Delta \beta}$, with $\hat{x}_T = \sum \hat{x}_k$ held fixed, is

$$\mathbf{\Delta T} = [\mathbf{DW\Lambda W^tD^t}] \cdot \mathbf{\Delta\beta}$$

so

$$\mathbf{\Delta\beta} = [\mathbf{DW\Lambda W^tD^t}]^{-1} \cdot \mathbf{\Delta T} \tag{3.28}$$

which can be inverted to obtain $\mathbf{\Delta\beta}$ provided

$$\det[\mathbf{DW\Lambda W^tD^t}] \neq 0 \tag{3.29}$$

This condition depends not only on the path matrix $\mathbf{D}$ but also on the ME estimates $\hat{x}_1, \ldots, \hat{x}_\mathrm{k}$ because they determine $\mathbf{G}$ by Eq. (3.21) and the eigenvalue and eigenvector matrices $\mathbf{\Lambda}$ and $\mathbf{W}$ for Eq. (3.29).

For noise analysis, we want to relate all the $\Delta\hat{x}_j$ to the $\mathbf{\Delta\beta}_i$ by Eq. (3.10), holding $\hat{x}_T$ fixed. Direct differentiation yields

$$\frac{\partial \hat{x}_k}{\partial \beta_i} = \left[\left(\mathbf{X} - \frac{1}{\hat{x}^t}\mathbf{xx^t}\right)\mathbf{D^t}\right]_{ki} = [\mathbf{GD^t}]_{ki} \tag{3.30}$$

and so the matrix statement of

$$\Delta\hat{x}_k = \sum_i \frac{\partial \hat{x}_k}{\partial \beta_i}\Delta\beta_i, \qquad k = 1, \ldots, K \tag{3.31}$$

is

$$\mathbf{\Delta x} = [\mathbf{GD^t}]\mathbf{\Delta\beta} = [\mathbf{W\Lambda W^tD^t}] \cdot \mathbf{\Delta\beta} \tag{3.32}$$

$\mathbf{\Delta x}$ is related to $\mathbf{\Delta T}$ through Eqs. (3.28) and (3.32). Formally it is

$$\mathbf{\Delta x} = [\mathbf{W\Lambda W^tD^t}] \cdot [\mathbf{DW\Lambda W^tD^t}]^{-1} \cdot \mathbf{\Delta T} \tag{3.33}$$

where $[]^{-1}$ is the inverse of the I × I matrix inside.

Equation (3.33) is, in fact, a generalized inverse solution of $\mathbf{D\Delta x} = \mathbf{\Delta T}$. To see this, we employ the following theorem [3.4]:

The general solution of the consistent nonhomogeneous equation

$$\mathbf{Ax} = \mathbf{y}, \qquad \mathbf{A} \text{ nonsquare in general,} \tag{3.34}$$

is

$$\mathbf{x} = \mathbf{A^-y} + (\mathbf{U} - \mathbf{A^-A})\,\mathfrak{X} \tag{3.35}$$

where $\mathbf{U}$ = unit matrix

$$\mathbf{A^-} = \mathbf{VA^t}(\mathbf{AVA^t})^{-1} \tag{3.36a}$$

$\mathbf{V}$ = arbitrary matrix, such that

$$\text{Rank}\,(\mathbf{AVA^t}) = \text{Rank}\,(\mathbf{A}) \tag{3.36b}$$

$\mathfrak{X}$ = arbitrary vector (column matrix here)

Notice the similarity in form of $\mathbf{A}^-$ to Eq. (3.33). In fact, if we associate $\mathbf{Ax} = \mathbf{y}$ with $\mathbf{D\Delta}x = \mathbf{\Delta T}$ and let $\mathbf{V} = \mathbf{W\Lambda W}^t$, we have Eq. (3.33) as the *generalized inverse solution* of $\mathbf{D\Delta}x = \mathbf{\Delta T}$, with arbitrary column matrix $\mathscr{X} = 0$.

According to the rank equality Eq. (3.36), we require

$$\text{Rank } (\mathbf{DW\Lambda W^tD^t})_{I\times I} = \text{Rank } (\mathbf{D})_{I\times K} \tag{3.37}$$

Now, if $(\mathbf{DW\Lambda}^{1/2})_{I\times K}$ has Rank I, $\mathbf{DW\Lambda}^{1/2}(\mathbf{DW\Lambda}^{1/2})^t = (\mathbf{DW\Lambda W^tD^t})$ has rank I, too. For every well-chosen set of rays whose path lengths are the rows of $\mathbf{D}$, the paths are independent and Rank $(\mathbf{D}) = \text{I}$. Therefore, the condition Eq. (3.37) for the generalized inverse solution Eq. (3.33) to exist with a proper inverse matrix requires

$$\text{Rank } (\mathbf{DW\Lambda}^{1/2})_{I\times K} = \text{Rank } (\mathbf{D})_{I\times K} = \text{I} \tag{3.38}$$

This is a subtle condition to interpret for behavior as the integration procedure converges toward an ME solution because $\mathbf{W}$ and $\mathbf{\Lambda}$ at each stage depend on $\mathbf{x}$ at that time [see Eqs. (3.21) and (3.22)].

Experience from a number of diversified problems indicates that, for a number of rays I in the neighborhood of K/2 there are $E_o \le I$ nearly zero eigenvalues of $\mathbf{\Lambda}$ and the matrices look like

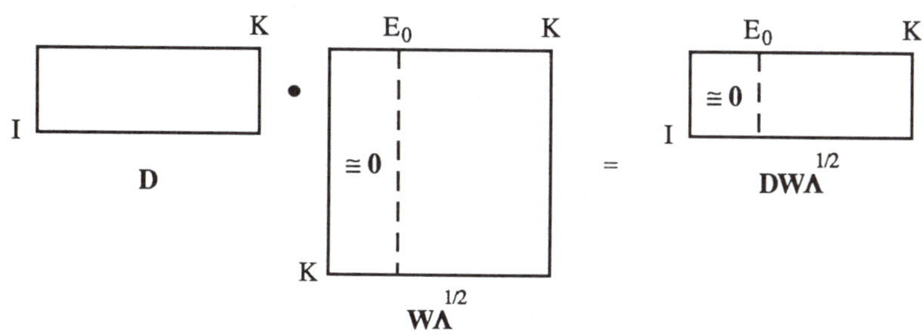

These matrices will yield essentially the same solution for $\mathbf{\Delta}x$ in Eq. (3.33) if we replace the nearly zero elements by zeros. Then, the rank of $\mathbf{DW\Lambda}^{1/2}$ will be I for $I < K/2$ and $K - E_o > I$. Equation (3.38) will be satisfied for the generalized inverse solution. As I increases for $I \gtrsim K/2$, E_o increases, and finally $K - E_o < I$. Then Rank $(\mathbf{DW\Lambda}^{1/2}) < I$, Eq. (3.38) is violated, and the generalized inverse solution Eq. (3.33) will not exist.

This reasoning shows that, for nearly zero elements in the previous matrices, the generalized inverse solution will behave qualitatively the same way and will become very ill-conditioned for $I \gtrsim K/2$ and $K - E_o < I$. A noise vector $\mathbf{\Delta x}$ computed from a noise image vector $\mathbf{\Delta T}$ would contain significant error in this situation.

There is another mathematical approach we can take to facilitate particular problems computationally. We can diagonalize the entire matrix in

Eq. (3.20) directly, without isolating the $\mathbf{\Lambda}$-eigenvalues of $\mathbf{G}$,

$$\mathbf{DGD^t} = \left[\mathbf{D}\left(\mathbf{X} - \frac{1}{\hat{x}_T}\mathbf{xx^t}\right)\mathbf{D^t}\right] = \mathbf{P}_{I\times I}\mathbf{\Gamma}_{I\times I}\mathbf{P^t}_{I\times I}, \qquad \text{with}$$
$$\mathbf{PP^t} = \mathbf{P^tP} = \mathbf{U} \tag{3.39}$$

In terms of these I × I matrices and new eigenvalue matrix $\mathbf{\Gamma}$ (diagonal), we can write the solutions Eqs. (3.28) and (3.33) as

$$\Delta\boldsymbol{\beta} = (\mathbf{P\Gamma P^t})^{-1}\,\Delta\mathbf{T} = (\mathbf{P\Gamma^{-1}P^t})\cdot\Delta\mathbf{T} \tag{3.40}$$

$$\Delta\mathbf{x} = \mathbf{W\Lambda W^tD^t}\cdot\mathbf{P\Gamma^{-1}P^t}\cdot\Delta\mathbf{T} \tag{3.41}$$

This solution may be justified as a special case of Eqs. (3.34) to (3.36) with $\mathbf{A} = \mathbf{P\Gamma P^t}$, $\mathbf{V} = \mathbf{P\Gamma^{-1}P^t}$. $(\mathbf{P\Gamma P^t})$ might be called an "abstract" path intersection matrix by analogy with $\mathbf{DXD^t}$, the path intersection matrix defined earlier, in terms of path matrix $\mathbf{D}$ and parameter matrix $\mathbf{X}$ (diagonal). The advantage of computing $\Delta\boldsymbol{\beta}$ in step (5) of each iteration of the ME algorithm is that we can tolerate more rays than are necessary ($I \gtrsim K/2$) and yet converge on a solution by contracting $(\mathbf{P\Gamma P^t})$ if the $\mathbf{\Gamma}$ matrix contains one or more very small eigenvalues. This is discussed in the next section.

The general requirements for a well-behaved ME solution are that the I ray paths are well chosen and that $I \lesssim K/2$, with K the number of object cells, so that both ranks in Eq. (3.38) are equal. They need not equal I but usually will. The Boltzmann ME geotomography technique is especially appropriate for vastly underdetermined problems, $I \ll K/2$.

3.5 A RELAXATION EIGENMATRIX CONTRACTION METHOD

To mitigate the unstable effect of zero or near-zero eigenvalues in the $(\mathbf{P\Gamma P^t})^{-1}$ matrix of Eq. (3.40), we have studied the effect of introducing a relaxation factor R into the $\mathbf{xx^t}$ term of Eq. (3.39). We rewrite Eq. (3.39) for a new matrix $\mathbf{G_R}$,

$$\mathbf{DG_RD^t} = \mathbf{D}\left(\mathbf{X} - \frac{1}{R\hat{x}_T}\mathbf{xx^t}\right)\mathbf{D^t} = \mathbf{P\Gamma P^t} \tag{3.42}$$

The idea is to decrease R from a large value to 1 as $\Delta\boldsymbol{\beta}$ and $\Delta\mathbf{T}$ both approach zero in the iterations of the algorithm toward convergence to the ME solution. We have found that the R-factor improves the numerical stability in many problems but does not avoid very small eigenvalues of the new $\mathbf{G_R}$ if the number of rays I is $\gtrsim K/2$, half the number of parameter cells in the object. But, if $K - E_o < I$, we can "stabilize" the solution with an eigenmatrix-contraction technique.

We can infer from the deductions of Lefeuvre and Delannoy [3.5] that the zero or near-zero eigenvalue part of $\mathbf{DGD}^t$ is irrelevant to an ME solution, and the smaller the eigenvalue of a vector the less that vector contributes to the solution. Following this philosophy each time we reach step (5) of the algorithm, we compute all I eigenvalues of $\mathbf{\Gamma}$ of Eq. (3.39) and the column eigenvectors $\mathbf{P}$ directly. Then, we delete those (I-m) columns of $\mathbf{P}$ with very small eigenvalues and contract $\mathbf{\Gamma}$ as well. This reduces Eq. (3.39) to (3.43), the tilde denoting a contracted matrix,

$$\tilde{\mathbf{P}}_{\mathrm{I}\times m}\tilde{\mathbf{\Gamma}}_{m\times m}\tilde{\mathbf{P}}^{t}_{m\times \mathrm{I}}, \qquad m < \mathrm{I} \tag{3.43}$$

and we use this to update $\boldsymbol{\beta}$ to $\boldsymbol{\beta} + \Delta\boldsymbol{\beta}$ in order to change $\hat{\mathbf{T}}$ to $\hat{\mathbf{T}} + \Delta\hat{\mathbf{T}} = \mathbf{T}$ (data) in Eq. (3.40),

$$\Delta\boldsymbol{\beta} = (\tilde{\mathbf{P}}\tilde{\mathbf{\Gamma}}\tilde{\mathbf{P}}^{t})^{-1}\cdot\Delta\mathbf{T} = \mathbf{A}^{-}\cdot\Delta\mathbf{T} \tag{3.44}$$

This solution is justified by Eqs. (3.34) to (3.36) if we associate $\mathbf{A} = \tilde{\mathbf{P}}\tilde{\mathbf{\Gamma}}\tilde{\mathbf{P}}^{t}$, $\mathbf{V} = \tilde{\mathbf{P}}\tilde{\mathbf{\Gamma}}^{-1}\tilde{\mathbf{P}}^{t}$ and assume that this $\mathbf{A}$ has rank I. $\mathbf{A}^{-}\cdot\mathbf{A} = \tilde{\mathbf{P}}\tilde{\mathbf{P}}^{t}$ in Eq. (3.35) but $\mathfrak{X}$ is taken to be zero.

The next time we reach step (5) of the algorithm this procedure is repeated in terms of the new **X**- and **x**-matrices in Eq. (3.42). The examples to follow indicate that it is generally necessary to have all the matrices uncontracted [i.e., m = I in Eq. (3.43)] to obtain the final convergence to the ME solution.

If we first employ the reduction factor R in Eq. (3.42) before contracting the matrices, and reduce R toward 1 as we increase m toward I as the iterations proceed, we have a highly effective technique for improving convergence to a final ME solution even when the number of rays exceeds K/2. Unfortunately, there appears to be no hard and fast rules for doing this; some trial and error is advisable for each problem.

3.6 SYNTHETIC DATA EXAMPLES [3.2]

We describe three two-dimensional examples of inversion of an underground parameter distribution from straight-ray synthetic data prescribed between two vertical boreholes in Fig. 3.1. The rectangular region is subdivided into square cells of unit length on a side.

Figure 3.3 shows an irregularly outlined region of $x = 3$ surrounded by $x = 4$ cells. The 20 rays employed are indicated by dotted lines and the ME values of the $\hat{x}_i$ are printed inside the 36 cells. These were obtained by starting with $\boldsymbol{\beta}^0 = \mathbf{0}$ in the algorithm and by solving Eq. (3.16) as matrix equation Eq. (3.28) for $\Delta\boldsymbol{\beta}$ each iteration by ordinary LU decomposition, Gaussian elimination, and back-substitution. Convergence was rapid even though one of the pivotal elements was very small, each iteration indicating a tiny eigenvalue of $\mathbf{\Lambda}$.

When we exercised the iteration algorithm by maintaining relaxation

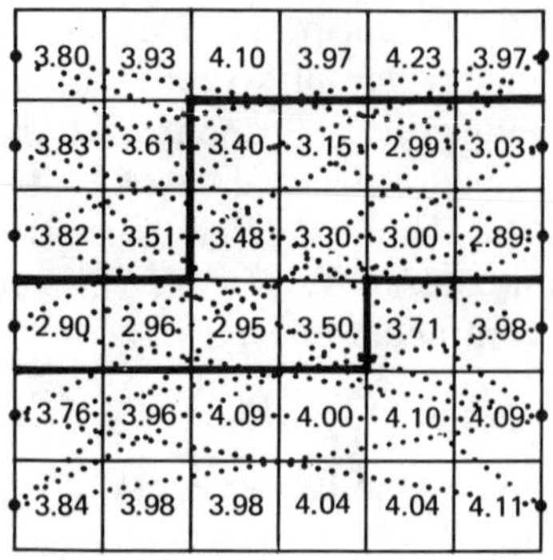

Figure 3.3 Boltzmann ME reconstruction of an irregularly outlined region of $x = 3$ cells surrounded by $x = 4$ cells. The ME-computed x_k values are indicated in the cells. From [3.2], Fig. 1, © 1981, IEEE.

factor R = 1 but omitting the small eigenvalue and its eigenvector from the matrices in Eq. (3.40) in *every* iteration, we converged to an ME solution with the wrong $\hat{\mathbf{T}} \neq \mathbf{T}$ data vector. Some of the $\hat{T}_i$ were correct but others were erroneous by a fraction $\cong 0.002$. This made the **x** distribution unacceptably different from the final ME distribution. That distribution was obtained by starting with R = 10 and reducing it to 1 on successive iterations, then finally adding the missing eigenvalue and its eigenvector back into the matrices of Eq. (3.40).

Resolution in Fig. 3.3 is rather good in that 8 of the 12 inner cells have their $\hat{x}_k$ within 0.2 of the true value 3, and 19 of the 24 outer cells have $\hat{x}_k$ within 0.2 of the correct 4.

Figure 3.4 illustrates the potential of the ME method for resolving contrasting local anomalies. The boldly outlined cell has an $x = 50$, while all the others have $x = 5$. The ME-obtained $\hat{x}_k$ are printed in the cells. The minimum number of rays utilized—without relaxation or matrix contraction—was the 25 shown dotted in Fig. 3.5. The ME solution *remained the same* when additional rays were added but the solution became more unstable, as indicated by increasing numbers of very small pivotal elements ($\lesssim 10^{-5}$) in the LU decomposition of Eq. (3.28). The CDC 7600 computer had sufficient word length of 14 significant decimal digits to invert Eq. (3.20) or (3.28) under these conditions but a computer of lesser word length would have suffered numerical overflow.

It seemed clear that about 25 rays was the maximum nonredundant number which would yield a good ME-Boltzmann solution. For the 25 rays indicated in the figure, there were no very small eigenvalues of $\mathbf{\Lambda}$. For more than 25 rays, the number of very small pivotal elements, equal to the number of small eigenvalues, was the excess of rays over 25—up to seven additional horizontal rays.

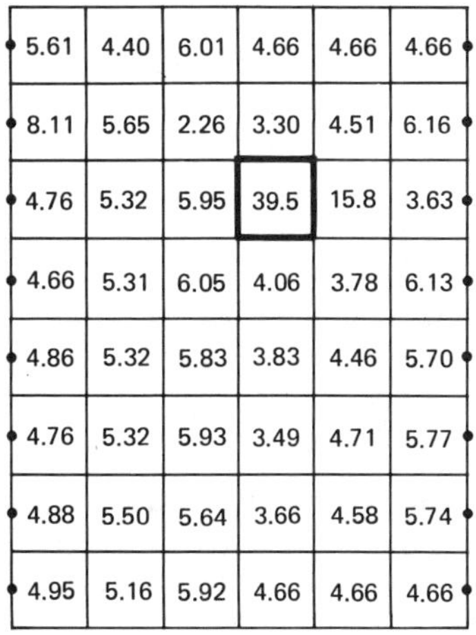

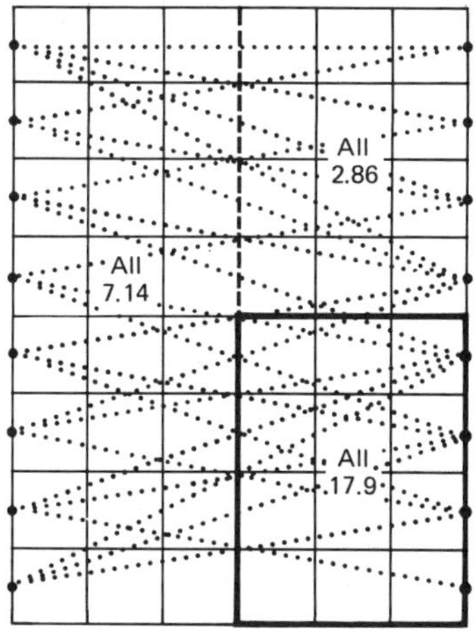

Figure 3.4 Boltzmann ME reconstruction of a two-dimensional region with a one-cell anomaly of $x = 50$ (heavily outlined box), surrounded by $x = 5$ cells. The rays are those of Fig. 3.5. The ME-computed x_k values are indicated in the cells. From [3.2], Fig. 2, © 1981, IEEE.

Figure 3.5 Boltzmann ME reconstruction of a two-dimensional region with a 12-cell anomaly of $x = 20$, surrounded by $x = 5$ cells. The ME-computed x_k values are indicated in the cells. From [3.2] Fig. 3, © 1981, IEEE.

Convergence to the correct ME solution was also rapid for more than 25 rays if relaxation factor R approached one from a large initial value in each step (5) of the iteration, with all the $\simeq 0$ eigenvalues and their eigenvectors omitted. This strategy worked even for 32 rays when seven or eight eigenvectors were removed in each iteration.

Figure 3.5 illustrates the resolution attainable from a localized contrast of $x = 20$ in the heavily outlined corner box of 12 cells, surrounded by $x = 5$ cells. The number of rays I was varied from 25 to 32 as in the preceding example. The ME $\hat{\mathbf{x}}$ distribution is indicated by the numbers within the three delineated regions shown. The solution was independent of whether ordinary LU decomposition and back substitution was used with R = 1 or the relaxation-contraction method with decreasing R. The latter method avoided the instabilities of the former.

The case of a 50:5 contrast between cells was also solved, with rapid convergence for 25 to 28 rays. The numbers replacing 7.14, 2.86, and 17.9 in Fig. 3.5 were 8.46, 15.39, and 46.5, respectively.

3.7 THE PROBLEM OF BENT RAYS

Generally, a ray traversing an inhomogeneous region like the one shown in Fig. 3.1 will change direction when crossing an interface in order to satisfy the boundary conditions (i.e., Snell's Law for refraction). The ME solution should incorporate these *interface conditions,* or *ray constraints,* as additional constraints on the problem.

The interface conditions merely reduce the number of possible macrostates of parameter distribution which also obey the measurement constraints. Entropy h is strictly a concave function of the $\{x_k\}$ and $x_T = \sum x_k$, and therefore, has no local minimum in $\mathbf{x}$ space. As we follow a trajectory continuously in $\mathbf{x} - \mathbf{D}$ space so as to maintain the measurement and interface constraints, h will maximize with respect to variations in either $\mathbf{x}$ or $\mathbf{D}$ which maintain the measurement constraints. "Virtual" variations in $\mathbf{D}$ around an ME point, whether or not they maintain the interface conditions[2], will upset the measurement constraints. These variations will change the $\{x_k\}$ in Eq. (3.10) and will decrease or increase h linearly. It is possible to have more than one ME solution in $\mathbf{x} - \mathbf{D}$ space, where all constraints are satisfied. The appropriate solution would be the one with the largest $h_{\max}$.

In the algorithm of Sec. 3.3 for converging on an ME solution, we must start with an initial guess of $\mathbf{D}$, $\mathbf{D}^{(0)}$, which usually will correspond to straight raypaths. In the iteration loop, we must update $\mathbf{D}$ each pass in order to iterate to the correct interface conditions. Usually, the best place to do this will be at step (1), after $\boldsymbol{\beta}$ determines the $\{\hat{x}_k\}$. Then, the raypaths can be changed in accord with the interface constraints and $\mathbf{D}$ re-evaluated. This will determine new $\{\hat{x}_k\}$, with the same $\boldsymbol{\beta}$, in Eq. (3.10). These, in turn, will change $\mathbf{D}$ again. After several cycles of $\mathbf{x} - \mathbf{D}$ iterations, the $\mathbf{D}$ should be sufficiently accurate at this stage of the computations to warrant proceeding through the remaining steps (2) to (5) of the algorithm. The updated $\mathbf{x}$ and $\mathbf{D}$ both update $\boldsymbol{\beta}$ in step (5).

3.8 STATISTICAL UNCERTAINTY AND ASSURANCE

This topic treats the uncertainty in the reconstructed object about its high-probability ME structure caused by the fact that many possible objects can produce the same measured image. We assume in this section that the image is noiseless. The data processing issues summarized in Fig. 1.9(a) are reproduced next in Fig. 3.6. The uncertainty in the true object caused by the many-to-one correspondence of objects to image is defined to be *statistical uncertainty*. The probability that the true object $\bar{x}^{\mathrm{ob}}$ (or one of its parameters x_k^{ob}) lies within a specified region $\delta\bar{x}^s$ (or δx_k^s) of its ME value is defined as the Assurance of that event.

First, we describe a high-probability region $\delta\bar{x}$ (with the superscript s

[2] Usually, for given $\{x_k\}$ there is only one set of rays which will travel from the transmitters to the various receivers and satisfy the interface constraints. If a ray is forbidden by the assumed $\{x_k\}$, it should be ignored in the process of iterating to an ME solution and perhaps reintroduced later when it is allowed by a later $\{x_k\}$ distribution.

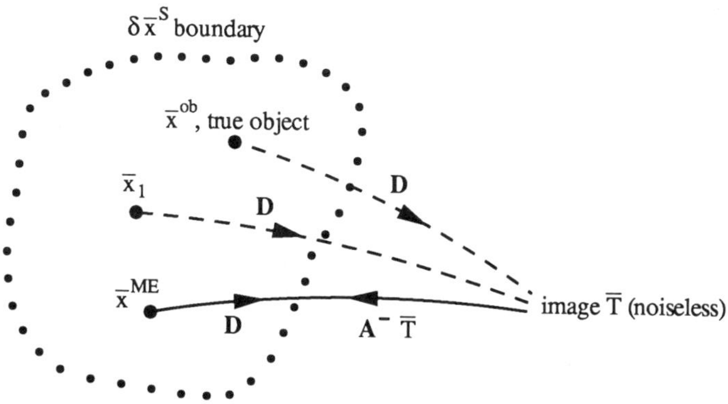

Figure 3.6 Object resolution from a noiseless image $\overline{T}$, reproduced from Fig. 1.9(a). The $(\delta\overline{x})^s$ object region around $\overline{x}^{ME}$ represents a high-probability region within which we might reasonably expect to find the true object $\overline{x}^{ob}$. The Assurance that $\overline{x}^{ob}$ lies within $(\delta\overline{x})^s$ is the probability that an $\overline{x}$ will lie within $(\delta\overline{x})^s$.

for statistical uncertainty implied) about $\overline{x}^{ME}$ by Jaynes's Concentration Theorem [3.6]. That Theorem expresses the probability of an object lying within an entropy interval δh of h_{max} and also satisfying the constraints, as

$$2N\ \delta h = \chi^2_{(K-I-1)}(s) \tag{3.45}$$

where

$$h = -\sum p_i \log p_i$$

Here, N is a finite number of trials of an experiment or elements in an object configuration (such as number of molecules, symbols in a possible message, elements of luminescence, or building blocks of a parameter distribution). X^2 is the Chi-squared distribution with (K − I − 1) degrees of freedom, K being the number of unknown parameters, I the number of constraints, and s the designated significance level for δh. This means the probability that an $\overline{x}$ object has entropy

$$H = N\ h(\overline{x}) = -N \sum_k \left(\frac{n_k}{N}\right) \log \left(\frac{n_k}{N}\right) \tag{3.46}$$

in the interval

$$h_{max} - \delta h \le h\ (\overline{x}) \le h_{max} \tag{3.47}$$

is precisely (1 − s), our *Assurance that the object lies within the* δh *region.*

For example, suppose we want δh to include 99 percent of all possible object configurations that satisfy the constraints. Suppose the number of degrees of freedom is (K − I − 1) = 20, with s = .01 so X^2_{20} (.99) = 37.6 from a statistical table. Then, δh = 18.8/N, which is $\ll 1$ if N is huge. We will say that the Assurance that the unknown object x lies within this δh interval is 99 percent.

The Concentration Theorem has involved this finite number N previously defined, which corresponds to quantizing a continuous parameter distribution to a discretization or discrimination level $\hat{dx} > 0$ in our preceding analysis of this chapter. The total amount of parameter $\hat{x}_T$ is $N\hat{dx}$, with $\hat{x}_k = \hat{n}_k \cdot \hat{dx}$ in the kth cell. It is important to realize that $\hat{dx}$ is not determined by measurement noise. $\hat{dx}$ was not relevant in obtaining the ME object $\bar{x}^{ME}$ because it was absorbed into the Lagrange multipliers. If we solve a problem according to the ME Eqs. (3.10) to (3.12) in one system of units and then change to a new system, such that the constraint relation $\mathbf{D}\ \mathbf{x} = \mathbf{T}$ is still satisfied but with different numerical values of $\mathbf{D}$, $\mathbf{x}$, and $\mathbf{T}$, the only effect in the ME solution is a multiplication of H by the scaling parameter for $\hat{x}_T$. We obtain the same ME solution as before but expressed in the new units.

It is easy to show that we cannot unambiguously study the distribution of $\bar{x}$ around $\bar{x}^{ME}$ merely by replacing N by $\hat{x}_T$ in Eqs. (3.45) and (3.46). Suppose we do so in the preceding example of the 99 percent confidence interval. Then, δh would become $18.8/\hat{x}_T$ and our confidence interval would be bounded by

$$h_{max} - \frac{18.8}{\hat{x}_T} \leq h(\bar{x}^{ob}) \leq h_{max} \tag{3.48}$$

If the units were changed to make $\hat{x}_T \rightarrow x'_T = r\hat{x}_T$ the 99 percent confidence interval would change to

$$h'_{max} - \frac{18.8}{(r\hat{x}_T)} \leq h'(\bar{x}^{ob}) \leq h'_{max} \tag{3.49}$$

This contradicts Eq. (3.48) because $h'_{max} = h_{max}$ in Eq. (3.46) $-$ $n_k/N = \hat{x}_k/\hat{x}_T$ is unchanged. This reasoning shows why the Concentration Theorem describes only the distribution of a *quantized* object $\bar{x}$ about its ME value.

The Assurance $\mathcal{A}$ that an object $\bar{x}$ satisfying the constraints lies close to the $\bar{x}^{ME}$ object implies nothing directly about the spread of a given x_k element about its x_k^{ME} value in the ensemble. To study such a parameter distribution we can utilize a Darwin-Fowler analysis [3.7]. This is outlined for quantized parameter distributions in Appendix C. The Darwin-Fowler analysis is formulated in terms of integral numbers of building blocks $\hat{dx}$ of a strictly positive (or negative) real object parameter such as density, photon intensity, or power. It is not strictly appropriate for Boltzmann analysis of field spectra in frequency or wavenumber space. Fields seldom have positive (or negative) semidefinite real and imaginary components.

The Darwin-Fowler analysis yields statistical moments of x_k about x_k^{ME} which approach the moments of a Poisson (or modified Bernoulli) distribution as $N \rightarrow \infty$

$$p\left(n_k = \frac{x_k}{\hat{dx}}\right) = \langle n_k \rangle^{n_k} \frac{e^{-\langle n_k \rangle}}{n_k!}, \qquad \langle n_k \rangle = n_k^{\mathrm{ME}} \tag{3.50}$$

This is appropriate for large $N \gg n_k$. The precise behavior with N can be estimated from the exact expressions for the second and third moments

$$\langle (n_k - \langle n_k \rangle)^2 \rangle = \langle n_k \rangle^2 \left[\frac{1}{\langle n_k \rangle} - \frac{1}{N}\right] \tag{3.51}$$

$$\langle (n_k - \langle n_k \rangle)^3 \rangle = \langle n_k \rangle^3 \left[\frac{1}{\langle n_k \rangle} - \frac{1}{N}\right]\left[\frac{1}{\langle n_k \rangle} - \frac{2}{N}\right] \tag{3.52}$$

The variance of $x_k = n_k \, \hat{dx}$ from Eq. (3.51) slightly less than

$$\langle (x_k - \langle x_k \rangle)^2 \rangle = \langle x_k \rangle \, \hat{dx} \tag{3.53}$$

The standard deviation $\sigma(x_k)$ due to the statistical uncertainty of x_k about $\langle x_k \rangle = x_k^{\mathrm{ME}}$ is the square root of Eq. (3.53),

$$\frac{\sigma(x_k)}{x_k^{\mathrm{ME}}} = \sqrt{\frac{\hat{dx}}{x_k^{\mathrm{ME}}}} \tag{3.54}$$

This valuable formula indicates that the fractional statistical uncertainty in a larger ME parameter is relatively smaller, neglecting any noise in image $\overline{\mathrm{T}}$.

Analogous to the Assurance $\mathcal{A}$ that an object $\overline{x}$ lies close to its ME value $\overline{x}^{\mathrm{ME}}$ according to Eq. (3.45), we define the Assurance that the true (unknown) value of x_k, x_k^{ob}, lies within an interval $x_{k1} \leq x_k^{\mathrm{ME}} \leq x_{k2}$ as the area of the Darwin-Fowler distribution in that interval

$$\mathcal{A}(x_{k1}, x_{k2}) = \sum_{n_{k1}}^{n_{k2}} p\left(n_k = \frac{x_k}{\hat{dx}}\right) \tag{3.55}$$

where $p(n_k)$ is often well estimated by Eq. (3.50).

The qualitative effect of $\hat{dx}$ on the Assurance $\mathcal{A}$ that an object parameter x_i^{ob} will lie within a specified interval about x_i^{ME} is demonstrated in Fig. 3.7. Small $\hat{dx}$ implies a finely resolved system, with higher probability that x_i^{ob} lies within the interval shown. Larger $\hat{dx}$ means the ensemble of objects all satisfying the constraints will be spread further away from x_i^{ME}, decreasing the probability that x_i^{ob} will lie within the same interval.

Clearly $\hat{dx}$ must be estimated realistically for meaningful Assurance statements that the true x_i^{ob} lies within a specified interval.

We now study the resolution of an object as it is discretized into successively more cells K and probed with an increasing number of rays I so as to maintain the ratio I/K. We let $\hat{dx}$ be fixed. For relatively few cells K, the amount of parameter x_k in cell k will be relatively high and the spread

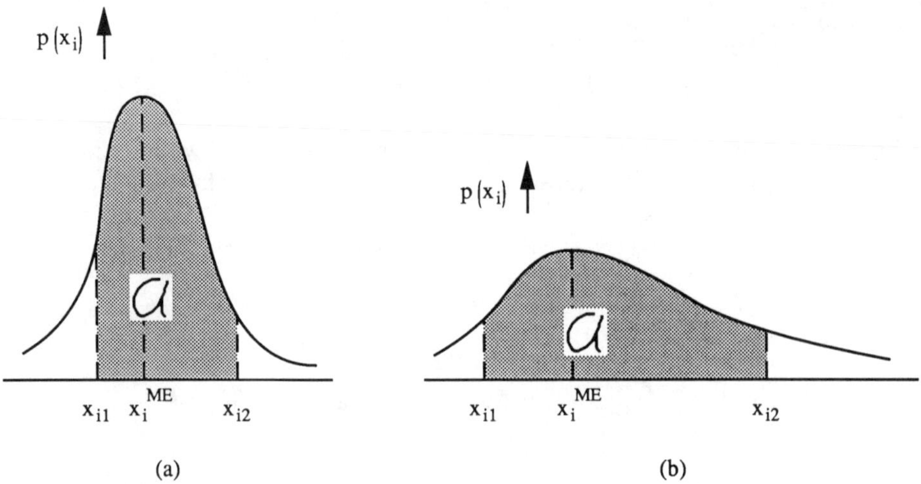

Figure 3.7 Qualitative dependence of Assurance $\mathscr{A}$ that x_i^{ob} will lie between x_{i1} and x_{i2}. Small $\hat{dx}$ implies a "tight" distribution of x_i about x_i^{ME}; large $\hat{dx}$, a "broad" distribution.

of possible x_k about x_k^{ME} will be relatively small according to Eq. (3.54). Thus, the resolution will be high in each cell but poor between cells because the discontinuities between neighbors will be rather large. If we increase the number of cells, with smaller x in each, the *intra*cell spread in parameter will increase according to Eq. (3.54) but the *inter*cell discontinuities will be smaller.

These considerations indicate one would not want to discretize an object to the point where the important cells were just a few $\hat{dx}$ dense.

3.9 STATISTICAL UNCERTAINTY WITH NOISE; CONFIDENCE

The more practical situation is now discussed of image noise due to measurement noise or external disturbances superimposed on the statistical uncertainty of the data processing previously discussed. The situation is depicted in Fig. 1.9(b) and reproduced in Fig. 3.8. The point $\overline{\text{T}}^{\text{ref}}$ represents a convenient image reference (not necessarily the point of zero noise), about which the image noise vector $(\delta\overline{\text{T}})^n$ describes a known probability region. $\overline{\text{T}}^{\text{ref}}$ yields the ME reference object $\overline{x}^{\text{ref}}$, and each image $(\delta\overline{\text{T}})^n$ produces a noisy ME object $(\delta\overline{x})^n$ away from the reference. Superimposed on each ME object, $\overline{x}^{\text{ref}} + (\delta\overline{x})^n$ is the independent statistical uncertainty described in Sec. 3.8.

We define the Confidence that the true object $\overline{x}^{ob}$ lies within a region shown in Fig. 3.8 as the probability that a point generated in object space by ME inversion of a noisy image and then perturbed by statistical uncertainty falls within that region.

With the algorithmic procedure of Sec. 3.3 we can in principle, solve for the ME $\bar{x}^{\text{ref}}$ from $\bar{\text{T}}^{\text{ref}}$ and any ME $\bar{x}^{\text{ME}} = \bar{x}^{\text{ref}} + (\delta\bar{x})^n$ mapped from $\bar{\text{T}} = \bar{\text{T}}^{\text{ref}} + (\delta\bar{\text{T}})^n$. For small $(\delta\bar{\text{T}})^n$ Eq. (3.33) can be used to obtain the $(\delta\bar{x})^n$. That formula was derived to give the change $\mathbf{\Delta x}$ in the ME object necessary to change $\bar{\text{T}}^{\text{ME}}$ by $\mathbf{\Delta T}$ and thus bring $\bar{\text{T}}^{\text{ME}}$ closer to $\bar{\text{T}}$ measured. However, $\mathbf{\Delta T}$ can just as well be a change due to noise or uncertainty, in which case $\mathbf{\Delta x}$ is the associated uncertainty in the object ($\hat{x}_{\text{T}}$ fixed, recall). Writing $\mathbf{G} = \mathbf{W\Lambda W^t}$ from Eq. (3.27) we have

$$\begin{aligned}(\delta\bar{x})^n \text{ from } \bar{x}^{ref} &\approx \mathbf{G}(\bar{x}^{ref}, \bar{\text{T}}^{ref})\mathbf{D^t}[\mathbf{DGD^t}]^{-1} \cdot (\delta\bar{\text{T}})^n \\ &\approx \mathbf{M}^-\ (\bar{x}^{ref}, \bar{\text{T}}^{ref}) \cdot (\delta\bar{\text{T}})^n\end{aligned} \tag{3.56}$$

$\mathbf{M}^-$ being the generalized inverse matrix. This can be used to obtain a first approximation to the noise δx_i in the ME x_i^{ME} in any cell i due to the measurement error $\delta\bar{\text{T}}$. The variance is

$$\langle[(\delta x_i)^n]^2\rangle \approx \sum_{j,k} \mathbf{M}_{ij}^- \ \mathbf{M}_{ik}^- \ \langle\delta\text{T}_j\ \delta\text{T}_k\rangle, \qquad \text{measurement noise} \tag{3.57}$$

To this is added the contribution to the variance due to the independent statistical uncertainty $(\delta x_i)^s$ about each $x_i^{\text{ME}} = x_i^{\text{ref}} + (\delta x_i)^n$. That variance

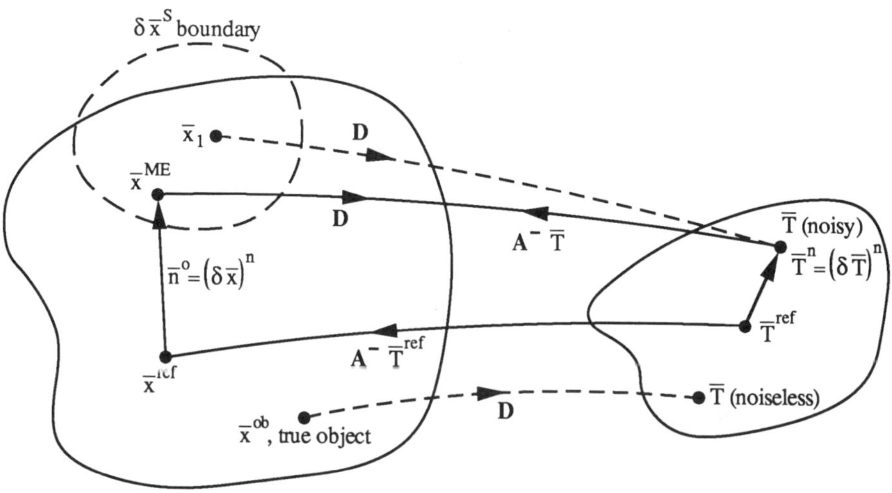

Figure 3.8 Object resolution with image noise present as introduced in Fig. 1.9(b) is shown with some notational changes. Image noise $\bar{\text{T}}^n$ is now called $(\delta\bar{\text{T}})^n$; object noise $\bar{\text{n}}^{\text{o}}$ is indicated by $(\delta\bar{x})^n$. $\bar{\text{T}}^{\text{ref}}$ is a reference image which yields the ME reference object $\bar{\text{x}}^{\text{ref}}$. Each noisy image $\bar{\text{T}}^{\text{ref}} + (\delta\bar{\text{T}})^n$ yields an ME noisy object $\bar{\text{x}}^{\text{ref}} + (\delta\bar{x})^n$. Superimposed on the $(\delta\bar{x})^n$ measurement noise is the independent object statistical uncertainty $(\delta\bar{x})^{\text{s}}$.

is Eq. (3.53). Therefore, the total variance in object parameter x_i about its x_i^{ref} value is approximately

$$\langle(\delta x_i)^2\rangle \approx \sum_{j,k} \mathbf{M}_{\bar{ij}}\ \mathbf{M}_{\bar{ik}}\ \langle\delta \mathrm{T}_j\ \delta \mathrm{T}_k\rangle + x_i^{ref}\ \hat{d}x \tag{3.58}$$

The Confidence $\mathscr{C}_i$ that a true object parameter x_i^{ob} is within c_i of its reference x_i^{ref} value would be computed generally as

$$\mathscr{C}_i\,(c_i) = \text{Prob}[|\ x_i - x_i^{ref}\ | \le c_i], \qquad x_i = x_i^{ref} + (\delta x_i)^n + (\delta x_i)^s \tag{3.59}$$

due to both image noise and statistical uncertainty.

REFERENCES

[3.1] P.M. Morse and H. Feshbach, *Methods of Theoretical Physics, Part 1,* Sec. 8.3, New York: McGraw Hill, 1953.

[3.2] R.M. Bevensee, Solution of underdetermined electromagnetic and seismic problems by the maximum entropy method, IEEE Trans. Ant. Prop., AP-29, March, 1981, pp. 271–274.

[3.3] E.T. Jaynes, Prior probabilities, IEEE Trans. Systems Sci. and Cyber., SSC-4, Sept., 1968, pp. 227–241.

[3.4] C.R. Rao and S.K. Mitra, *Generalized Inverse of Matrices and Its Applications,* New York: John Wiley, 1971. Theorems 2.1.1.1, p. 20, and 2.3.1, p. 23.

[3.5] F. Lefeuvre and C. Delannoy, Analysis of random electromagnetic wave field by a maximum entropy method, (in English), *Annales Des Telecommunications, 34,* March-April, 1979, pp. 204–213.

[3.6] E.T. Jaynes, On the rationale of maximum entropy methods, Proc. IEEE, 70, Sept., 1982, pp. 939–952.

[3.7] R.B. Lindsay and H. Margenau, *Foundations of Physics,* New York: Dover, 1957, pp. 252–279.

4

Maximum Entropy Tomography (Boltzmann Method)

4.1 INTRODUCTION

This chapter considers the problem of reconstructing a two-dimensional object from a number of transmitted, intersecting ray data by a different ME procedure than we employed in Chap. 3. The iterative procedure that follows does not require computation of logarithms or exponentials and is particularly appropriate for inverse reconstruction from a small number of views. With few object views, iterative methods often perform better than Fourier-space inversion and convolution backprojection.

The ME method is described later for resolution of an electron beam density, and could also be applied to plasmas under magnetic confinement. The method could be adapted to invert ray data transmitted through a cross section of human bone and tissue cells and yield CAT (Computer-aided Tomography) scans of the body. CAT has been successfully applied to nondestructive evaluation of objects for minute flaws and imperfections.

We show how to extend the method to include prior information and allow for missing data. Prior information is especially effective for object enhancement when only limited-view data are available.

The method can be generalized straightforwardly for reconstruction of three-dimensional bodies, for which its relatively small storage requirements are a decided advantage.

4.2 DESCRIPTION OF THE GENERAL PROBLEM

The general problem is to obtain an estimate of the density distribution of an object in a cross-sectional slice from rays probing the slice. We illustrate the ME method of doing this for a relativistic electron beam which is made to generate light at every point in its cross section proportional to the electron density. The light comes from nitrogen molecules struck by the electrons. The rays of light transmitted in various directions through the cross section furnish the data for resolving the instantaneous electron density in that cross section.

Figure 4.1 shows the essence of the experimental arrangement. The electron beam emerging perpendicularly from the plane of the paper at some

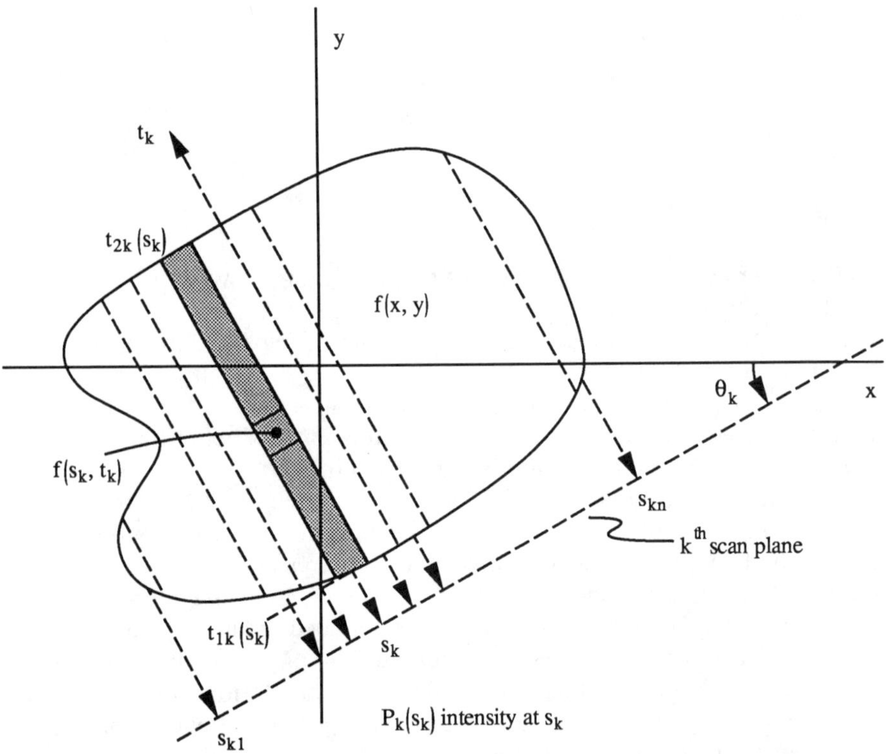

Figure 4.1 An electron beam emerging perpendicular to the paper has a cross-sectional density distribution $f(x, y)$ in the plane of the paper. This is sampled on a number of scan planes, the kth of which is shown at angle θ_k with the x axis. At a typical point s_k in this plane, the total light intensity $P_k(s_k)$ is proportional to the total electron density $\int f(x, y)dt = \int f(s_k, t_k)dt_k$ across the shaded strip shown. The tomogram of $P_k(s_k)$ across the kth scan plane, thereby, samples $f(x, y)$ in strips.

time T has a cross-sectional density distribution $f(x, y)$ within the irregular contour in the plane of the paper. This is essentially constant during the short time of interrogation by a sequence of optical scans, recorded on film during a 0.5 ns (nanoseconds; 1 ns $= 10^{-9}$ sec) interval. The kth optical scan plane is shown at angle θ_k to the x axis. At each point s_k on this plane, an optical fiber sequencing system[1] records the light coming from one of the strips of electrons, which are colliding with nitrogen atoms in the path of the beam. The total light intensity P_k at point s_k for the shaded strip shown is

$$P_k(s_k) = \int_{t_{1k}(s_k)}^{t_{2k}(s_k)} dt_k \, f(s_k, t_k) \tag{4.1}$$

ignoring the constant of proportionality between the light intensity $f(s_k, t_k)$ and the electron density at (s_k, t_k). (This is established separately.)

In order to relate the integration coordinates s_k, t_k on a strip to the reference coordinates x, y we note the $s_k - t_k$ coordinate system is just rotated counterclockwise by θ_k radians from the $x - y$ coordinate system. Ergo,

$$s_k = x \cos \theta_k + y \sin \theta_k \tag{4.2}$$

$$t_k = -x \sin \theta_k + y \cos \theta_k \tag{4.3}$$

and the inverse relations

$$x = s_k \cos \theta_k - t_k \sin \theta_k \tag{4.4}$$

$$y = s_k \sin \theta_k + t_k \cos \theta_k \tag{4.5}$$

Equation (4.1) furnishes one measurement constraint on $f(x, y)$, and all such measurements over all s_k on the kth scan plane and over all such scan planes furnish the set of constraints for the ME estimate, $f(x, y)^{\mathrm{ME}}$.

4.3 MATHEMATICAL PROCEDURE

The entropy of the object (electron beam cross section at a given time) is the Shannon form

$$h = -\int dx \int dy \, f(x, y) \log[f(x, y)] \tag{4.6}$$

[1] In more detail, the light from each strip in Fig. 4.1 falls on one member of an array of parallel optical fibers. The fibers transmit light signals sequentially across each scan plane, one plane after the other, to a special "streak" camera. This camera converts the signals to proportional electrical pulses which are recorded. The entire process of recording all the scan plane data is so rapid that an ME tomographic image of the cross-sectional electron beam density has about a 1 ns resolution. A highly relativistic electron beam would travel nearly 1 ft in 1 ns.

from Eq. (2.27) with neglect of the term $+ \int dx \int dy\, f(x, y) \log E$, $E = \int dx \int dy\, f(x, y)$. This term would introduce a trivial factor to the right side of Eq. (4.11).

The integrals are discretized into cells of sides Δx, Δy in the data processing according to the resolution of the measurement apparatus. However, the data processing shall be described in terms of integrals. Constraint Eq. (4.1) is multiplied by a Lagrange multiplier $\beta_k(s_k)$, a *continuous* function of s_k, and integrated over s_k

$$\int ds_k \int \beta_k(s_k) \left[\int_{t_{1k}(s_k)}^{t_{2k}(s_k)} dt_k\, f(s_k, t_k) - P_k(s_k) \right] = 0 \tag{4.7}$$

The objective function F is h plus the sum of all the constraints Eq. (4.7) over the various scan planes $k = 1, \ldots, N$

$$F = h + \sum_{k=1}^{N} \int ds_k \int \beta_k(s_k) \left[\int_{t_{1k}}^{t_{2k}} dt_k\, f(s_k, t_k) - P_k(s_k) \right] \tag{4.8}$$

This is made an extremum (maximum for h) by setting every variation δF equal to zero for an arbitrary $\delta f(x, y)$. The result is straightforwardly obtained

$$\begin{aligned} \delta F = 0 = &- \int dx \int dy\, [1 + \log f(x, y)]\, \delta f(x, y) \\ &+ \sum_k \int_{\text{scan plane } k} ds_k\, \beta_k(s_k) \int_{t_{1k}}^{t_{2k}} dt_k\, \delta f(s_k, t_k) \end{aligned} \tag{4.9}$$

Note that

$$\delta f(s_k, t_k) = \delta f(x = s_k \cos\theta_k - t_k \sin\theta_k,\ y = s_k \sin\theta_k + t_k \cos\theta_k)$$

is taken at a particular (x, y) related to (s_k, t_k) as indicated.

In order for Eq. (4.9) to be zero for all arbitrary localized variations $\delta f(x, y)$ at various (x, y) in the beam, the integrand of Eq. (4.9) must be zero

$$0 = -\{1 + \log[f(x, y)]\} + \sum_k \beta_k(s_k = x\cos\theta_k + y \sin\theta_k) \tag{4.10}$$

This readily yields

$$f(x, y) = e^{-1} \exp\left[\sum_k \beta_k(s_k = x\cos\theta_k + y\sin\theta_k) \right] \tag{4.11}$$

$e^{-1} = .36788\ldots$ is just a scale factor that can be absorbed into the constant of proportionality between electron density f and light intensity $P_k(s_k)$. So we ignore it.

Equation (4.11) says that $f(x, y)$ at (x, y) is determined by the sum of the Lagrange multiplier functions, each evaluated at a coordinate s_k on the

kth scan plane determined by the x and y in question. [The corresponding, quite different geotomographic expression is Eq. (3.10).] Since the exponential form in Eq. (4.11) is not advantageous in this problem, we define new Lagrange functions η_k as

$$\eta_k\,(z) = \exp\,[\beta_k\,(z)] \tag{4.12}$$

With these we rewrite Eq. (4.11) as

$$f(x, y) = \prod_l \eta_l\,[s_l(x, y)], \qquad s_l(x, y) = x \cos \theta_l + y \sin \theta_l \tag{4.13}$$

s_l applies to the lth scan plane.

4.4 EVALUATION OF THE LAGRANGIAN FUNCTIONS $\eta_l(s_l)$

We substitute Eq. (4.13) into constraint Eq. (4.1)

$$\begin{aligned} \mathrm{P}_k(s_k) &= \int dt_k\, f(s_k, t_k) = \int_{t_{1k}(s_k)}^{t_{2k}(s_k)} dt_k\, f[x(s_k, t_k), y(s_k, t_k)] \\ &= \int_{t_{1k}}^{t_{2k}} dt_k \prod_l \eta_l[s_l(x, y)] \end{aligned} \tag{4.14}$$

Now it is computationally expedient to have the integrand entirely in terms of coordinates s_k and t_k. So, we express $s_l\,(x, y)$ of h_l in terms of s_k and t_k using Eqs. (4.4) and (4.5)

$$\begin{aligned} s_l(x, y) &= x \cos \theta_l + y \sin \theta_l = x(s_k, t_k) \cos \theta_l + y(s_k, t_k) \sin \theta_l \\ &= (s_k \cos \theta_k - t_k \sin \theta_k) \cos \theta_l + (s_k \sin \theta_k + t_k \cos \theta_k) \sin \theta_l \\ &= s_k \cos (\theta_l - \theta_k) + t_k \sin (\theta_l - \theta_k) \end{aligned} \tag{4.15}$$

With this constraint Eq. (4.14) is rewritten

$$\mathrm{P}_k(s_k) = \int_{t_{1k}(s_k)}^{t_{2k}(s_k)} dt_k \prod_{l=1}^{\mathrm{K}} \eta_l\,[s_k \cos (\theta_l - \theta_k) + t_k \sin (0_l - \theta_k)], \qquad k = 1, \ldots, N \tag{4.16}$$

s_k sweeps across scan plane k. For data processing with the iterative algorithm [4.1] we separate η_k on the right and have

$$\mathrm{P}_k(s_k) = \eta_k(s_k) \int_{t_{1k}(s_k)}^{t_{2k}(s_k)} dt_k \prod_{l \neq k} \eta_l[s_k \cos (\theta_l - \theta_k) + t_k \sin (\theta_l - \theta_k)], \qquad k = 1, \ldots, N \tag{4.17}$$

Equation (4.17) is the basis of the ME iterative routine to converge on all the self-consistent $\eta_k(z)$ functions. When those are found, the relative electron density function $f(x, y)$ is known by Eq. (4.13). $P_k(s_k)$ is the light intensity measured at point s_k on the kth scan plane.

The Gauss-Seidel scheme for updating each $\eta_k(s_k)$ from $\eta_k^{(n)}$ to $\eta_k^{(n+1)}$ in an ME computer code developed for this tomography [4.1] is simply

$$\eta_k^{(n+1)}(s_k) = \frac{P_k(s_k)}{\int_{t_{1k}(s_k)}^{t_{2k}(s_k)} dt_k \prod_{l \neq k} \eta_l^{(n)}\left[s_k \cos(\theta_l - \theta_k) + t_k \sin(\theta_l - \theta_k)\right]} \tag{4.18}$$

for all s_k along the kth scan plane and for the other planes. The initial $\eta_k^{(0)}$ are taken all equal to an estimated value.

4.5 EXAMPLE OF AN ELECTRON BEAM RECONSTRUCTION

Figure 4.2 shows a time sequence of six representative cross-sectional objects (electron density profiles or tomograms) and their ME-reconstructed two-dimensional plots of the electron density of a 4.5 MeV (million electron volts), 60 kA (kiloampere), 30 ns electron beam pulse from the ETA (Experimental Test Accelerator) at Lawrence Livermore National Laboratory [4.2]. The mountainous figure for each plot shows the variation of electron density (proportional to the mountain height) over the cross-sectional plane. The data were obtained from only three scan planes, at 0°, 61°, and 117° relative to the x axis of Fig. 4.1. Scan plane data for each cross-sectional image were taken during a snapshot interval of nearly 0.5 ns and recorded so rapidly that the reconstructed ME object or plot has a temporal resolution of $\cong$ 1 ns.

A total of 70 snapshots were reconstructed during the 50 ns interval depicted in Fig. 4.2. A nice feature of the ME method is the reproduction of zero portions of the tomograms without "ringing" artifacts.

4.6 DATA PREPROCESSING FOR THE BEAM

If the data are carefully taken, the total energy E_k in the kth scan plane should be the same as any other E_l because

$$\begin{aligned} E_k &= \int ds_k\, P_k(s_k) = \int ds_k \int dt_k\, f(s_k, t_k) \\ E_l &= \int ds_l\, P_l(s_l) = \int ds_l \int dt_l\, f(s_l, t_l) = E_k \end{aligned} \tag{4.19}$$

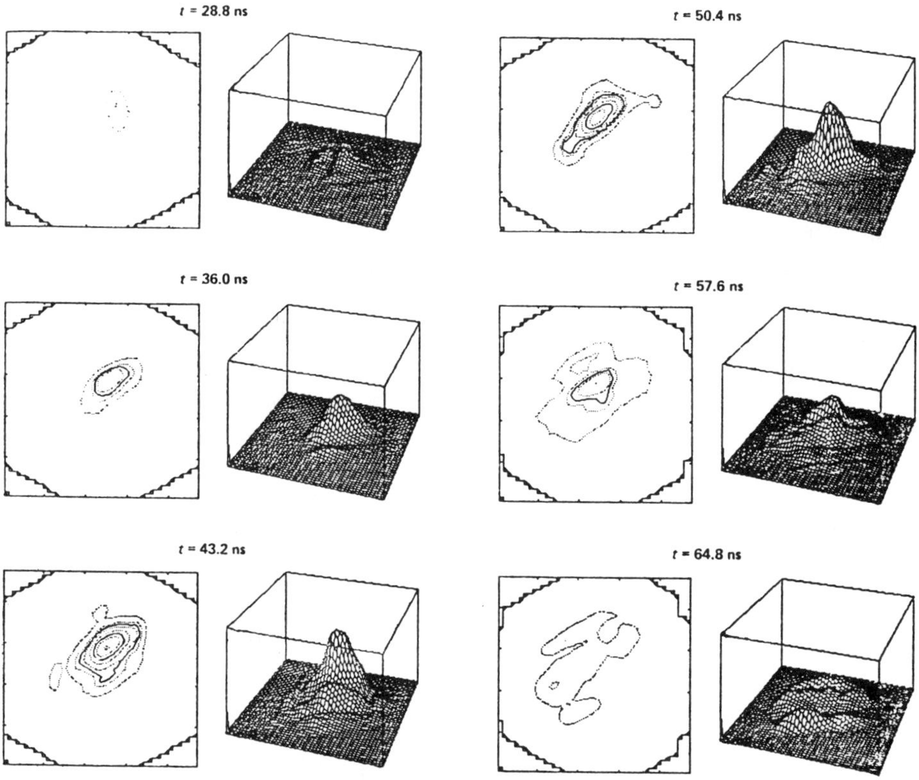

Figure 4.2 A sequence of six ME-reconstructed "snapshots" of cross sections of a 4.5 MeV, 60 kA electron beam pulse. The figure to the left of each snapshot shows contours of constant electron density. From [4.2] Fig. 6.

both measure the total energy $\mathrm{E} = \int dx \int dy\, f(x, y)$ according to Eq. (4.1) and Fig. 4.1. The ME technique does presume consistent data, and any deviations among the E_k would have to be attributed to noise and/or instrumentation error. To minimize these effects the computer code for this problem finds E from $\mathrm{P}_1\ (s_1)$ of the first scan according to Eq. (4.19) and normalizes each $\mathrm{P}_k(s_k)$ of subsequent scans of the same cross section to yield the same E. Perhaps a better value of E would be obtained from an average of all the scans.

The second consistency condition concerns the centers of "masses" (image intensities on the scans), defined as

$$(s_{om})_k = \mathrm{E}^{-1} \int ds_k\, s_k\, \mathrm{P}_k(s_k) \qquad \text{for the } k\text{th scan} \tag{4.20}$$

The projections of three or more of these $(s_{om})_k$ along perpendiculars to the scan planes should ideally all meet at the same point in the beam contour—

the center of mass of the cross section. In practice, a "compromise" center of mass is found which is a minimum distance, in a mean-square sense, from all the projection lines. Each $P_k(s_k)$ pattern is then slid sideways on its scan plane (maintaining the same energy E) until its center of mass projection line passes through the compromise point.

We could obtain an estimate of energy E of the cross section according to these considerations and add more constraints to force the various energies E_k measured on the scan planes in Eq. (4.19) to be E. However, this ploy does not obviate the effects of noise and/or error and the data processing does not essentially change: Eq (4.17) would still determine all the $\eta_k(s_k)$ functions and $f(x, y)$ by Eq. (4.13)! To appreciate this, let us add n more constraints of the form Eq. (4.19) to the objective function Eq. (4.8). We add

$$\sum_{k=1}^{n} \gamma_k \left[\int ds_k \int dt_k\, f(s_k, t_k) - \mathrm{E} \right] \tag{4.21}$$

in which the new Lagrange multipliers γ_k are scalars rather than functions of s. This would change Eq. (4.11) to

$$f(x, y) = e^{-1} \exp \left[\sum_k \beta_k(s_k = x \cos \theta_k + y \sin \theta_k) + \gamma_k \right] \tag{4.22}$$

(where e^{-1} can be absorbed as before). Only one overall amplitude $\exp(\sum \gamma_k)$ has been added to $f(x, y)$ regardless of how the γ_k are evaluated individually. The data processing proceeds as before, by Eq. (4.17) for each k plane, in terms of the same functions η_l. The solution for $f(x, y)^{\mathrm{ME}}$ insures the satisfaction of Eq. (4.19) for the various scan planes just as if the data had been preprocessed to scale all the $P_k(s_k)$.

4.7 COMPUTATIONAL STABILITY AND NOISE

The tomographic data analysis of Sec. 4.3 is formally similar to the geotomographic analysis of a discrete-cell region in Sec. 3.2. The discretized form of $f(x, y)$ is $f(s_k, t_k)$ and this can be labeled f_κ. The datum $P_k(s_k)$ of the kth scan at s_k can be ordered by the single parameter ι. If all tomographic cells have the same size their entropy can be written from Eq. (4.6) as

$$h = -\sum_{\kappa}^{K} f_\kappa \log (f_\kappa) \tag{4.23}$$

and the constraint equations Eq. (4.1) ordered by $k - s_k$ can be rewritten

$$\mathrm{P}_\iota = \sum_\kappa \mathrm{D}_{\iota\kappa} f_\kappa \tag{4.24}$$

where $D_{\iota\kappa}$ is the distance traveled by the $\iota \sim k - s_k$ ray through the $s_k - t_k$ cell. In these forms, Eq. (4.23) is equivalent to H of Eq. (3.8), with n_k measuring the amount of material, x_k, corresponding to f_κ in the kth cell. Equation (4.24) is equivalent to constraint equation Eq. (3.3), with P_ι replacing T_i.

In view of the close analogy, we expect the considerations of noise and stability would apply to the ME tomography data processing as well. Equation (3.33) written with the tomographic parameters describes the noise $\mathbf{\Delta f}$ in the cells in terms of measurement noise $\mathbf{\Delta P}$ as

$$\mathbf{\Delta f} = [\mathbf{W\Lambda W^tD^t}] \cdot [\mathbf{DW\Lambda W^tD^t}]^{-1} \cdot \mathbf{\Delta P} \tag{4.25}$$

in which the inverse matrix exists, if the situation is sufficiently underdetermined: I sufficiently less than $\simeq$ K/2. $\mathbf{W}$ and $\mathbf{\Lambda}$ are the eigenvector and eigenvalue matrices of the "material density" matrix $\mathbf{G}$

$$\mathbf{G} = \mathbf{f_D} - \frac{1}{\hat{f}_T}\mathbf{ff^t} \tag{4.26}$$

with $f_D = \text{diag}(f_1, \ldots, f_\kappa)$, $\mathbf{f} = \text{col}(f_1, \ldots, f_\kappa)$.

To estimate the underdetermined nature of typical tomographic problems, assume the object cross section is circular and interrogated by n strips in each scan plane. Let the cross section be gridded by $n \times n$ cells. Since each ray of a scan is perpendicular to a diameter each scan has n rays. With $N = 3$ scan planes as a practical number, the total number of rays I is $3n$. The total number of cells K is $\simeq \pi(n/2)^2$. According to the criterion of Sec. 3.4, the ME inversion procedure might become ill-conditioned when I/(K/2) $\cong$ 1. Here I/(K/2) is $\cong 24/(\pi n)$, which is <1 for $n \geqslant 8$ strips per scan plane. Such discretization is usually realized in practice; the electron beam resolution described earlier processed nearly 100 strips per scan plane.

The analogy between the geotomographic analysis of Chap. 3 and the tomographic analysis suggests Eq. (3.53) for estimating the statistical uncertainty in a "point" density $f(x, y)$. With the expected value $\langle f \rangle = f^{ME}$ we have

$$\langle [f(x, y) - f(x, y)^{ME}]^2 \rangle^{1/2} = [\hat{d}f \cdot f(x, y)^{ME}]^{1/2} \tag{4.27}$$

The left side is the standard deviation of the intensity at (x, y) measured over all members of the f ensemble which satisfy the measurement constraints and $\hat{d}f$ is the level of resolution in f over the object. $\hat{d}f$ is usually $\ll$ all $f(x, y)^{ME}$ of the object.

A noise analysis is reported in [4.1] for the resolution of a solid cylinder inside a hollow cylinder from data taken on 5 scan planes. Random noise was introduced into the various P_k (s_k) scan values at 5 percent noise levels. The ME algorithm proved to be "fairly stable" at these levels. The fluctuations in the reconstructed source (object) suffered an average noise amplitude of about 10 percent.

4.8 THE TREATMENT OF PRIOR INFORMATION AND MISSING DATA

N. J. Dusaussoy [4.3] and [4.4] has modified the preceding methodology by introducing a discrimination function after Eq. (1.48), which includes a prior-information estimate $f_p(x, y)$ of the unknown density function $f(x, y)$. When this is minimized subject to the constraints of the measurements along the various scan planes and allowing for missing data, if any, the resultant ME-Boltzmann $f(x, y)^{\mathrm{ME}}$ is obtained. The exponentiated Lagrange multipliers $\{\eta_k\}$ for the discrete strips of data are iterated to convergence by the Gauss-Seidel method and Dusaussoy proves that the procedure converges to the unique $f(x, y)^{\mathrm{ME}}$ distribution which minimizes the discrimination function subject to the constraints. The proof does not depend on the continuity of $f(x, y)$.

Reconstructions from limited-angle scan data are improved by prior information. The extended MENT algorithm, based on iterations, usually performs better than conventional algebraic reconstruction techniques and convolutional backprojection. Such iterative algorithms as MENT often resolve objects better than transform algorithms when data are missing.

Instead of minimizing the discrimination function we shall maximize its negative, h_{D},

$$h_{\mathrm{D}} = -\int dx \int dy\, f(x, y) \log\left[\frac{f(x, y)}{ef_p(x, y)}\right] \tag{4.28}$$

with properties described in Sec. 1.7 relative to h_p of Eq. (1.47). h_{D} has only one maximum, at all $f(x, y) = f_p(x, y)$ because of $e = 2.71828\ldots$ in the denominator. When h of Eq. (4.6) is replaced in Eq. (4.8) by h_{D} the variational procedure yields Eq. (4.11) again but with e^{-1} replaced by $f_p(x, y)$. Equation (4.13) is replaced by

$$f(x, y) = f_p(x, y) \prod_l \eta_l[s_l(x, y)] \tag{4.29}$$

Dusaussoy applies the Gauss-Seidel method to iterate to the final values of the $\eta_k\,(s_k)$ of Eq. (4.18), with $f_p[x(s_k, t_k), y(s_k, t_k)]$ in the integrand, as follows. If a $\mathrm{P}_k\,(s_k) = 0$ (an unlikely zero-datum) $\eta_k^{(0)}\,(s_k)$ and successive $\eta_k^{(n)}\,(s_k)$ remain zero. Otherwise, all $\eta_l^{(0)}\,(s_l)$ are set to 1. If a datum $\mathrm{P}_m(s_m)$ is missing, the corresponding $\eta_m^{(n)}\,(s_m)$ is not updated and remains 1. After iterating cyclically all the $\eta_k(s_k)$ to convergence and obtaining all the $f(x, y)^{\mathrm{ME}} \geq 0$ for all (x, y) by Eq. (4.28) it follows that $f(x, y) = 0$ only on those strip(s) for which $\eta_k\,(s_k) = 0$.

Execution time is measured by the number of additions, $p\mathrm{N}_s\mathrm{N}_p^2$, for each iteration cycle of Eq. (4.18) and the number of multiplications plus divisions, $q\mathrm{N}_s\mathrm{N}_p^2$. Here N_p is the number of scan planes, and N_s is the number of strips on each scan plane. p and q depend on the method of discretizing the integrands in Eq. (4.18) and on the required precision of the numerical integration. p and q can be $\approx$ 1000 for the adaptive trapezoidal rule.

Dusaussoy observes that the intensity fluctuations due to statistical uncertainty in synthetic-data examples appear from experiments to be proportional to object density. Defining a fluctuation by the left side of Eq. (4.27), this implies

$$\langle [f(x, y) - f(x, y)^{\mathrm{ME}}]^2 \rangle^{1/2} \propto f(x, y) \tag{4.30}$$

This is qualitatively similar to the prediction of Eq. (4.27). The coefficient of proportionality in Eq. (4.30) depends on the Lagrange multipliers, which depend on the data in an involved way. Without knowing the details of this involvement, we can only say that theoretical considerations support the observations that Eq. (4.30) is generally true.

4.9 AN EXAMPLE OF RECONSTRUCTION WITHOUT AND WITH PRIOR INFORMATION

Figure 4.3 shows a cross section of a simulated object of varying radial density measured by $f(x, y)$ and containing two defects, a rectangle of density $f = 3.0$ in the upper right portion of the outer ring ($f = 5.0$) and a smaller ellipse of $f = 0.4$ in the lower right portion of the inner ring ($f = 1.0$). The two rings shown represent a typical cross section of a conical structure.

There is also a known opaque obstruction of $f = \infty$ on the x axis which intercepts some of the probing rays falling on 16 scan planes, equally spaced between $\theta = 0°$ and 168.75°. Three of these planes are shown, at angles of 0°, 78.75°, and 168.75°. Each scan plane of width 2 intercepts 128 parallel rays, so there are $16 \times 128 = 2048$ data points. 307 of these are missing because of the obstruction.

Equation (4.1) represents the datum $P_k(s_k)$ for ray s_k on scan plane k. If the rays were X rays, $f(s_k, t_k)$ would be the *logarithm* of the attenuation per unit length, proportional to object density. $P_k(s_k)$ would be the *logarithm* of the total attenuation suffered by ray s_k. In this synthetic data example no noise is assumed in the ray data or in the data processing. Differences between the true and reconstructed objects are due to statistical uncertainty, with negligible numerical roundoff error.

The object in Fig. 4.3 is reconstructed within $128 \times 128 = 16384$ pixels using the extended MENT algorithm, first without prior information and with $f_p(x, y)$ taken to be the average $f(x, y)$ computed from Eq. (4.19). Missing data are treated according to the remarks following Eq. (4.29). The results are shown in Fig. 4.4, (a) being the direct reconstruction and (b) showing the "difference" object. That object is obtained by subtracting the

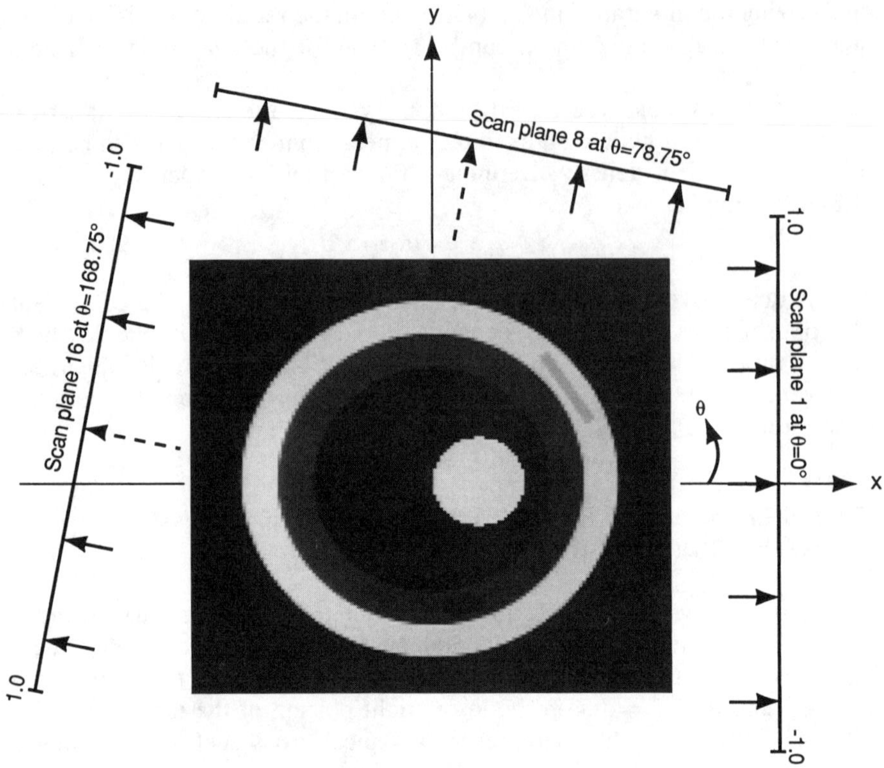

Figure 4.3 Cross section of an object with varying radial density measured by $f(x, y)$, including two defects of $f = 3.0$ and 0.4 and an opaque obstruction of $f = \infty$. Ray data are collected at 128 points on each of 16 scan planes at angular separations of 11.25° over $0 \leq \theta \leq 168.75°$. Three of the scan planes are shown at angles of 0°, 78.75°, and 168.75°. From [4.3] Fig. 6.3.

object without defects (and without the obstruction) from the reconstruction in (a). The rectangular defect contrasts rather well with the pattern of "ghost" defects around the outer ring, but the small elliptical defect is barely discernable in the inner ring. The obstruction shows up because its prior f_p is an average f which is not updated in the Gauss-Seidel iterations. The magnitude of the difference between this prior f and the true f for no obstruction shows up as the difference obstruction of Fig. 4.4(b).

The object of Fig. 4.3 reconstructed with its prior $f_p(x, y)$ representing the object without defects and without the obstruction is shown in Fig. 4.5, (a) being the direct reconstruction and (b) the "difference" object obtained in the same way as with no prior information. Now the difference object is nearly zero except in the clearly defined defect regions. The obstruction

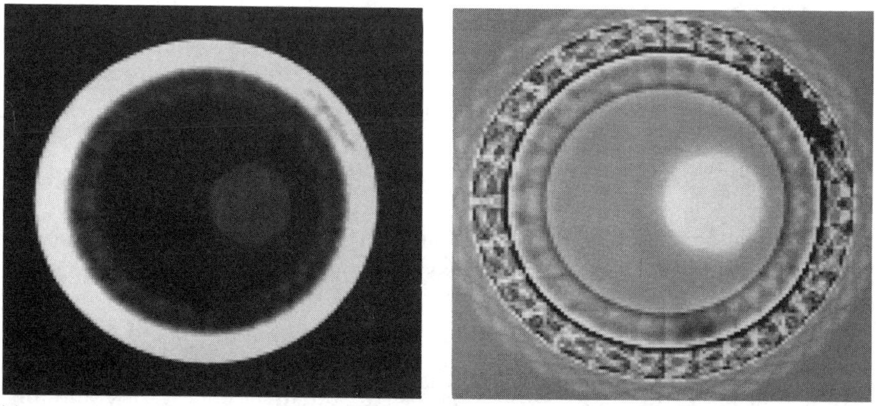

(a) (b)

Figure 4.4 (a) Reconstruction of the object of Fig. 4.3 on 128 × 128 pixels by extended MENT, with prior $f_p(x, y)$ taken to be the average f over the object. (b) The "difference" object, obtained by subtracting the true object without defects (and without the obstruction) from the reconstruction (a). The defect in each ring is discernable against the circular pattern of "ghost" defects. From [4.3] Fig. 6.21.

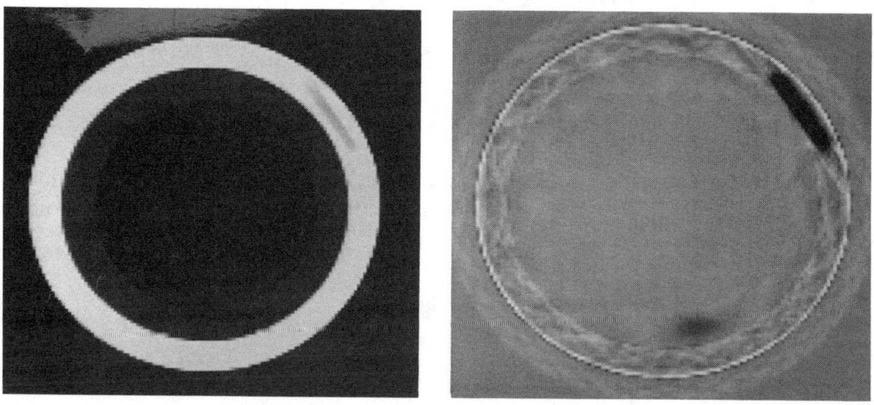

(a) (b)

Figure 4.5 (a) Reconstruction of the object of Fig. 4.3 on 128 × 128 pixels by extended MENT, with prior $f_p(x, y)$ representing the object without defects and without the obstruction. (b) The "difference" object, obtained by subtracting the true object without defects (and without the obstruction) from the reconstruction (a). Now the two defects are more obvious. From [4.3] Fig. 6.22.

does not appear because it was correctly represented by its prior in Fig. 4.5(a).

In reconstructions of this sort, a figure of merit is the discrete residual norm, defined for the nth iteration of the Gauss-Seidel method as

$$E_{res}^{(n)} = \left\{ \frac{1}{M} \sum_k \left[P_k^{(n)}(s_k) - P_k^{meas}(s_k) \right]^2 \right\}^{1/2} \tag{4.31}$$

where the sum is taken over M data points on all the scan planes. $E_{res}^{(n)}$ usually decreases rapidly at first with increasing n before leveling off to a nearly constant value near zero.

In order to obtain good contrast quality like that of Figs. 4.4 and 4.5, the contrast differences should be closely proportional to object "density" $f(x, y)$ and larger than any variations due to statistical uncertainty and any noise fluctuations.

Extended MENT outperforms algebraic reconstruction techniques and convolution backprojection for problems of this sort, particularly when data are missing.

REFERENCES

[4.1] G. Minerbo, MENT: A maximum entropy algorithm for reconstructing a source from projection data, Computer Graphics and Image Processing, 10, 1979, pp. 48–68.

[4.2] H.A. Koehler, B.A. Jacoby, and M. Nelson, Time-resolved tomographic images of a relativistic electron beam, Proc. of the Society of Photo-Optical Instrumentation Engineers, Aug. 19–24, 1984.

[4.3] N.J. Dusaussoy, Maximum entropy type algorithms for image reconstruction from projections, dissertation for the degree of Doctor of Philosophy in Electrical Engineering, University of Delaware, December, 1988.

[4.4] N.J. Dusaussoy and I.E. Abdou, The extended MENT algorithm: A maximum entropy type algorithm using prior knowledge for computerized tomography, IEEE Trans. Signal Proc., *39*, pp. 1164–1180, May, 1991.

5

Maximum Entropy Spectral Analysis (Gibbs's Method)

5.1 INTRODUCTION

The spectral analysis of this chapter estimates either the frequency content of a finite temporal signal or ensemble of such signals or the wavenumber content (directions of travel of the waves) in a bounded spatial-temporal signal or ensemble of them. Examples are a speech pattern characterized by frequency peaks and an earthquake seismic wave, which is analyzed by an array of sensors to determine the direction of travel from the epicenter. Spectral analysis is an involved subject because the signals are often of short duration and many methods are available for empirically extrapolating the properties of the long-time signals. Classical spectral estimation concentrates on direct analysis of measured waveforms. Modern methods of analysis seek to obtain a model of the process which generates a measured signal or ensemble of them. The model parameters determine the spectral properties of these signals.

A powerful, popular, and efficient method of spectral estimation is based on the so-called autoregressive (AR) or all-pole model of waveform generation. The formula for the spectrum of a time-limited waveform, in terms of the parameters of the AR model evaluated from autocorrelation data, is supported by maximum entropy considerations. The primary purpose of this chapter is to demonstrate this support, and to derive the analogous formula for a wavenumber spectrum in terms of spatial-temporal autocorrelation data. Our secondary purpose is to put the AR method of spectral estimation in perspective with other contemporary methods.

We characterize a temporal signal in general as the output of a system characterized by a finite-order linear ordinary differential or difference equation and driven by noise, noise plus sinusoids or sinusoids. Such a system has poles and/or zeros in its frequency-domain input-output transfer function. The spatial array waves are presumed propagating in definite directions as plane waves but with the temporal character of the signal just described.

We state the basic Fourier transform relations and the formulas for autocorrelation function of a temporal signal and its transform, the power spectrum. For a sampled temporal signal $f(t)$ and its Fourier transform $F(\omega)$ the Nyquist or sampling relations show that the time resolution Δt is related to maximum frequency $\omega_m = 2\pi f_m$, and, *independently*, the frequency resolution $\Delta\omega$ is determined by the measurement time T. The discrete fast Fourier transforms are based on these sampling relations. Several classic spectral estimators are described.

The famous Burg ME power spectrum in the frequency domain is derived from a probability distribution for the discrete temporal raw data string which maximizes the Shannon entropy subject to measurement autocorrelations. We follow the derivation of McDonough. The Burg ME method extrapolates the unknown autocorrelation functions by the same recursion relation as an autoregressive (AR) filter. Since a power spectrum is a function of the autocorrelation functions, the AR filter output power spectrum is the Burg ME spectrum. The Burg power spectrum is generated at the filter output by white noise input.

Then we follow the work of Rietsch and derive the Burg power spectrum as the "expected" spectrum of a probability distribution for the power spectral amplitudes themselves. This distribution maximizes the Shannon entropy subject to autocorrelation data constraints.

The autocorrelation data processing in the time-frequency domains carries over to the space-wavenumber domains when we make the proper correspondences between time and space parameters and between frequency and wavenumber parameters. The wavenumber spectrum at a chosen frequency as the transform of a defined power spectrum is analogous to the earlier frequency power spectrum as the transform of its autocorrelation function. The space-time autocorrelation function is generally complex but the wavenumber spectrum has the Burg form.

In order to put the ME and AR methods in perspective, we first summarize the salient features of autoregressive moving average (ARMA) or pole-zero modeling and moving average (MA) or all-zero modeling. We then discuss the "Prony Method" of pole-zero modeling as an alternative to ME or AR all-pole modeling. The Prony Method is flexible enough to represent any process of data generation with a model of general pole-zero structure. The so-called "Square" or Determined Prony method is applied to estimated autocorrelation data with a sum-of-exponentials representation. The amplitudes (residues) and complex poles in this representation are found. The

Overdetermined Prony method fits the estimated data in a least-squares sense. Both versions of the Prony method may be applied to raw data as well. The "Overdetermined" Prony method (with the Square or Determined Prony method as a special case) is equivalent to Makhoul's method of covariance modeling.

We summarize Marple's study of the effect of noise and signal-to-noise ratio (SNR) on the ME superresolution of two coherent sinusoidal signals very close in frequency. Resolution is generally better than by conventional windowing methods, for SNR values down to -20 dB and improves with the number of autocorrelation data. These results apply to wavenumber resolution of coherent sources very close in directional angle with receiver noise present.

Wavenumber resolution of incoherent sources as a function of SNR can be improved over that of conventional methods by removing the known noise component from the autocorrelation data before data processing by the Gibbs-Burg method.

5.2 TIME SERIES ANALYSIS (TIME AND FREQUENCY DOMAINS); BASIC RELATIONS

A temporal signal $f(t)$ is often analyzed to determine its spectral characteristics, which are signatures of the generating process. For example, the complex resonant frequencies and their amplitudes in an EEG (electroencephalogram) furnish information about the owner's health. The complex resonant frequency pattern in a radar return signal helps to describe the general features of an unknown target.

In time-frequency, or t-ω, analysis the fundamental relations are embodied in the Fourier transform pair

$$F(\omega) = \int_0^\infty f(t)e^{-j\omega t}\,dt, \qquad -\infty < \omega = 2\pi f < \infty \tag{5.1}$$

$$f(t) = \frac{1}{2\pi}\int_{-\infty}^{\infty} F(\omega)\, e^{+j\omega t}d\omega, \qquad 0 \le t < \infty \tag{5.2}$$

$F(\omega)$ is the spectral complex amplitude at circular frequency $\omega = 2\pi f$ in radians per second. f is the frequency in Hertz, not to be confused with $f(t)$, the real-time signal, which is assumed to start at $t = 0$. Because $f(t)$ is real during the measurement interval T, we have the spectral relation $F(-\omega) = F^*(\omega)$, the asterisk denoting complex conjugate.

Related parameters are the autocorrelation function $r(\tau)$, defined formally as

$$r(\tau) = \lim_{T\to\infty} \frac{1}{T}\int_0^T f(t)\,f(t + \tau)dt, \qquad r(-\tau) = r(\tau) \tag{5.3}$$

and the power spectrum S(ω),

$$S(\omega) = \int_{-\infty}^{\infty} r(\tau)e^{-j\omega\tau}d\tau \tag{5.4}$$

In practice, one measures t-domain data over a finite time interval (0,T) with equipment which is effectively limited in bandwidth. The mathematical consequence of not knowing the temporal waveform $f(t)$ for $t > \mathrm{T}$ is that T becomes a fundamental period, with $\Delta\omega = 2\pi/\mathrm{T}$ the fundamental circular frequency. It is the limit of resolution in frequency[1] which one can reliably attain. The spectrum extends over many multiples of Δω, or harmonics of the fundamental frequency.

However, there is a limit to the highest positive frequency $\omega_M = \mathrm{M}\Delta\omega$ which one can specify in the waveform $f(t)$ from samples at equal increments of time Δt over the interval (0,T). One way to understand this is by duality: the transforms Eqs. (5.1) and (5.2) are interchanged if we substitute $\mathrm{F}(\omega) \to \sqrt{2\pi}f(t)$, $f(t) \to \mathrm{F}(-\omega)/\sqrt{2\pi}$, $\omega \to t$, $t \to \omega$. Hence, the sampling relation $\Delta\omega = 2\pi/\mathrm{T}$ must have a dual. This is obtained by replacing Δω with Δt and T with $2\omega_M$, the entire meaningful frequency interval from $-\omega_M$ to $+\omega_M$. This yields $\Delta t = 2\pi/2\omega_M = \pi/\omega_M$ usually written $\Delta t = 1/(2f_M)$. Any equipment which linearly processes such a waveform at intervals Δt should have a bandwidth of ω_M or less to avoid distortion. Saying it another way, if the bandwidth is $(-\omega_M, +\omega_M)$ one can resolve $f(t)$ exactly at intervals no finer than $\Delta t = \pi/\omega_M$. Analogously, one can resolve F(ω) exactly at intervals no finer than $\Delta\omega = 2\pi/\mathrm{T}$.

These mentioned relations $\Delta\omega = 2\pi/\mathrm{T}$ and $\Delta \mathrm{t} = \pi/\omega_M$ are known as the Nyquist sampling criteria. It is important that they are *independent of each other* in practice [5.1]. An $f(t)$-waveform of duration T is specified uniquely during that interval by frequencies at multiples of Δω, but the signal bandwidth $(-\omega_M, +\omega_M)$ is determined by how finely we resolve the waveform; that is, by Δt. Conversely, a waveform of known bandwidth is uniquely specified at time intervals of Δt, independently of the duration T.

Suppose the bandwidth were limited to $(-\omega_M, \omega_M)$ but that raw data samples of $f(t_n = n\Delta t)$ were available at intervals of $\Delta t = 1/(2f_M)$ over the entire range $(0 \leq t \leq \infty)$. This implies perfect frequency resolution: $\Delta f = 0$. Then, it can be shown [5.2, p. 36] that $f(t)$ at any t is computed exactly by interpolation as

$$f(t) = \sum_{n=0}^{\infty} f(t_n)\,\frac{\sin[\omega_\mathrm{M}(t - t_n)]}{\omega_\mathrm{M}(t - t_n)}, \qquad t_n = n\Delta t \tag{5.5}$$

In practice only a finite number of samples (N + 1) are available, and if we

[1] The word *frequency* usually denotes a circular frequency $\omega = 2\pi f$ rather than f in Hertz. The context should avoid confusion.

interpolate in the interval $(0, N\Delta t)$ by truncating the sum to $n = N$ in Eq. (5.5), we will not generally represent $f(t)$ correctly *between* its samples.

The Nyquist criteria imply the same number of frequency intervals resolved in $(-\omega_M, +\omega_M)$ as time intervals in (0,T)

$$\frac{2\omega_M}{\Delta\omega} = \frac{T}{\Delta t} = 2f_M T = N \qquad \left(f_M = \frac{\omega_M}{2\pi}\right) \tag{5.6}$$

This property is utilized in a fast Fourier transform (FFT) algorithm.

The Uncertainty Principle of signal resolution is sometimes written as $\Delta\omega\Delta T \geq 1$, or $\Delta f \geq 1/(2\pi\Delta T)$. This refers to the accuracy Δf with which one may measure a frequency in a specified measurement time ΔT. The "ΔT" is our signal duration T, so the Principle is consistent with one of the Nyquist relations. It says nothing about the $\omega_M \Delta t = \pi$ relation.

These Nyquist criteria are emphasized because they will be relevant to the ME data processing to superresolve spectra at frequency intervals $\Delta\omega < 2\pi/T$.

According to the criteria, we should replace Eqs. (5.1) and (5.2) in practical data analysis by the discrete forms

$$F_m = F(m\Delta\omega) = \sum_{n=0}^{N} f_n\, e^{-jm\Delta\omega n\Delta t}\, \Delta t, \qquad T = N\,\Delta t \tag{5.7}$$

$$f_n = f(n\Delta t) = \frac{1}{2\pi} \sum_{m=-M}^{M} F_m e^{+jm\Delta\omega n\Delta t}\, \Delta\omega, \qquad \omega_M = M\,\Delta\omega \tag{5.8}$$

Here M, the number of positive (or negative) frequency components equals N/2. In an FFT algorithm, N is usually a power of 2, though it need not be [5.3].

In a finite raw data record of duration T we must estimate $r(\tau)$ of Eq. (5.3) as $\hat{r}(j\Delta\tau) = \hat{r}_j$, where $\Delta\tau = \Delta t$. Assuming N raw data samples over $0, \Delta t, \ldots, (N-1)\Delta t = T$ one good estimator to use is

$$\hat{r}_j = \frac{1}{N-j} \sum_{n=0}^{N-j-1} f_n f_{n+j} = \langle f_n f_{n+j} \rangle_T \tag{5.9}$$

Here, $\langle\ \rangle_T$ denotes an average over one data record. If this f-generating process is ergodic (equal $T = \infty$ average over one data record and ensemble average of records each T long) then, the ensemble-averaged $\langle \hat{r}_j \rangle_E$ is an unbiased estimator of the true r_j in the sense that [5.2, p. 147]

$$\langle \hat{r}_j \rangle_E = \langle \langle f_n\, f_{n+j} \rangle_T \rangle_E = r_j \tag{5.10}$$

If factor $(N - j)$ were replaced by N in Eq. (5.9) $\langle \hat{r}_j \rangle_E$ would be biased for finite N.

The estimator $\hat{S}(\omega)$ obtained by discretizing Eq. (5.4) and using the $\hat{r}_j$

of Eq. (5.9) is

$$\hat{S}_{AUTO}(\omega_m = m\Delta\omega) = \Delta t \sum_{n=0}^{2L} \hat{r}_n e^{-jm\Delta\omega n\Delta t} \tag{5.11}$$

(subscript AUTO means autocorrelation) where L is typically $\ll$ the number of raw data samples, N. This inequality offsets the greater statistical variance in those $\hat{r}_j$ of greater lag j. The ensemble average $\langle S_{AUTO}(\omega_M)\rangle_E$ is biased even though the $\langle \hat{r}_j\rangle_E$ are not.

To reduce this bias, each estimate $\hat{r}_n$ is "windowed" by multiplying it by a "window" factor in Eq. (5.11). The various types of windows are described in [5.2, Sec. 5.3].

The "periodogram" estimate, $\hat{S}_{PER}$ of S, based on Eq. (5.7) and raw data samples at 0, Δt, . . . , $N\Delta t$ in a data record is

$$\hat{S}_{PER}(\omega_m = m\Delta\omega) = \left\langle \frac{1}{(N+1)\Delta t} \mid F_m \mid^2 \right\rangle_E \tag{5.12}$$

averaged over an ensemble of data records. This is the original unmodified periodogram spectral estimate, which would yield inconsistent estimates if one were to ignore the ensemble-averaging operation. Various periodogram techniques have been developed to improve this spectral estimate [5.2, pp. 152–161].

5.3 SPECTRAL FREQUENCY RESOLUTION BY THE GIBBS-BURG METHOD (PROBABILITY DISTRIBUTION FOR THE TIME SERIES)

The Gibbs-Burg ME power spectrum $S(\omega)$ was first derived by J. P. Burg [5.2], who maximized the entropy proportional to $\int \log[S(\omega)]d\omega$ for a time-limited function $f(t)$ for which the known sequence of autocorrelation functions was extrapolated by one more value. We regard the following derivation of McDonough [5.4] as fundamental. In Sec. 5.4 the same spectrum is derived from a different Gibbsian point of view.

We assume the single raw data temporal sequence $\{f_i\}$ is processed at intervals $t_n = n\Delta t$ over an interval $(0, T_p)$, where $T_p = p\Delta t \leq T = N\Delta t$. Often it is instructive to study the behavior of the power spectrum as the order p of the process increases to its maximum of N, the number of measurement intervals. The fact that there are fewer $f_n \cdot f_{n+p}$-terms with which to compute r_p for larger p is another reason for choosing $p < N$, while evaluating the $\hat{r}_0, \ldots, \hat{r}_p$ over the entire measurement interval in Eq. (5.9).

When we introduce the multivariate probability $P(f_1, \ldots, f_N)$ and maximize the entropy $-\int df_1 \cdots \int df_N \, P \log P$ by the procedure in Appendix D with the Lagrange multipliers $\beta_0, \ldots, \beta_p$ (reals), we obtain a multivariate

Gaussian expression for P in which the (N + 1) × (N + 1) autocorrelation matrix $\mathbf{R}_N$ is the inverse of the (N + 1) × (N + 1) $\boldsymbol{\beta}$-matrix, which has a triangle of zeros in the upper right and lower left corners [see App. D, Eq. (D.9)]. The β_i's are chosen to make $r_0, \ldots, r_p$ agree with the autocorrelation data; the remaining $r_{p+1}, \ldots, r_N$ are ME estimates [5.5].

We can now proceed to derive the ME power spectrum in either of two ways. We can adjoin to the data $f_1, \ldots, f_{N+1}$ an extrapolated value $\tilde{f}_{N+2}$ (denoted by the tilde), append an additional $\tilde{r}_{N+1}$, enlarge $\mathbf{R}_N$ to $\mathbf{R}_{N+1}$ and use the multivariate Gaussian function for $P(f_1, \ldots, \tilde{f}_{N+2})$ in the procedure to follow in order to derive the ME power spectrum. Or we can avoid the large matrices involved and define a reduced data train $f_1, \ldots, f_{p+1} (p < N)$, append an extrapolated $\tilde{f}_{p+2}$, let $P(f_1, \ldots, \tilde{f}_{p+2})$ have the multivariate Gaussian form in terms of the smaller $(p + 2) \times (p + 2)$ matrix $\mathbf{R}_{p+1}$ of measured $r_0, \ldots, r_p$ and extrapolated $\tilde{r}_{p+1}$, and proceed next to maximize the entropy by proper choice of $\tilde{r}_{p+1}$. We will follow the latter procedure, because it can be iterated for successively larger $p = p_{\min}$, $p_{\min+1}, \ldots$, N by the Levinson algorithm [5.6], stopping at $p_{\text{final}} \leq N$ determined by a prediction error criterion, such as that of H. Akaike [5.6].

One should keep in mind when defining the "order" of a process that there is no need to pursue this iteration procedure; the full power of the ME method allows one to process directly the data autocorrelations $r_0, \ldots, r_q$ defined for any q over the full data string $f_1, \ldots, f_{N+1}$ and obtain a frequency spectrum in which the Lagrange multipliers weigh the $\{r_i\}$ according to their relevance [5.5].

We now proceed to analyze a raw data train $f_1, \ldots, f_{p+1}$ with the measured autocorrelations $r_0, \ldots, r_p$ defined in Eq. (5.9). We extend the sequence by Δt with a hypothetical datum $\tilde{f}_{p+2}$ at time $(p + 1)\Delta t$ and an unknown autocorrelation coefficient $\tilde{r}_{p+1}$. The tilde denotes these extrapolated quantities. With T_p increased to $(T_p + \Delta t)$ the spectral interval $\Delta\omega$ is slightly reduced and spectral resolution increased—but over the same bandwidth $(-\omega_M, +\omega_M)$ corresponding to Δt.

To adjoin $\tilde{f}_{p+2}$ to the temporal sequence in a maximum entropy way, we introduce the joint probability density P of an ensemble of possible sequences, a typical one that is denoted

$$P(f_1, f_2, \ldots, \tilde{f}_{p+1}, \tilde{f}_{p+2}) \tag{5.13}$$

This probability distribution has the Shannon entropy

$$h = -\int df_1 \cdots \int d\tilde{f}_{p+2} P(f_1, \ldots, \tilde{f}_{p+2}) \log P(f_1, \ldots, \tilde{f}_{p+2}) \tag{5.14}$$

In Appendix D, we derive the maximum entropy P subject to the constraints that the measured $r_0, \ldots, r_p$ agree with the ensemble averages $\langle r_0 \rangle, \ldots, \langle r_p \rangle$. These averages are taken over an ensemble of f-sequences, each of length $(p + 2)$, generated by P. The maximum-entropy P is

$$P(f_1, \ldots, f_{p+1}, \tilde{f}_{p+2})$$

$$= \left[(2\pi)^{(p+2)/2} \, \| \mathbf{R_{p+1}} \|^{1/2}\right]^{-1} \exp\left[-\frac{1}{2}\, \mathbf{f^t R_{p+1}^{-1} f}\right] \tag{5.15}$$

Here

$$\mathbf{f} = \text{column}(f_1, \ldots, \tilde{f}_{p+2}), \quad \mathbf{f^t} = \text{transposed } \mathbf{f} \tag{5.16}$$

$\| \mathbf{R}_{p+1} \|$ is the determinant of the correlation matrix with the Toeplitz form

$$\mathbf{R_{p+1}} = \begin{bmatrix} r_0 & r_1 & r_2 & \cdots & \cdots & r_p & \tilde{r}_{p+1} \\ r_1 & r_0 & r_1 & \cdots & \cdots & r_{p-1} & r_p \\ \vdots & \vdots & & & \vdots & \vdots & \vdots \\ \vdots & \vdots & \vdots & \cdots & r_0 & r_1 & r_2 \\ r_p & r_{p-1} & \cdots & \cdots & r_1 & r_0 & r_1 \\ \tilde{r}_{p+1} & r_p & \cdots & \cdots & r_2 & r_1 & r_0 \end{bmatrix} \tag{5.17}$$

Because Eq. (5.15) describes a train of raw data $(f_1, \ldots, \tilde{f}_{p+2})$ by a multivariate Gaussian probability function the ME procedure is strictly appropriate only for Gaussian random processes.

Substitution of Eq. (5.15) into Eq. (5.14) shows that entropy h is proportional to $\log\| R_{p+1} \|$ within an irrelevant constant, that is,

$$h = \kappa \log[\det(\mathbf{R_{p+1}})] \tag{5.18}$$

To maximize h we choose $\tilde{r}_{p+1}$ in the upper right and lower left corners of Eq. (5.17) to maximize the determinant. We differentiate it with respect to $\tilde{r}_{p+1}$ and set the result to zero. With the matrix identity

$$d[\det(\mathbf{M})] = \det(\mathbf{M})\cdot\text{Trace}(\mathbf{M^{-1}\, dM}), \qquad \text{Trace } \mathbf{A} = \sum A_{ii} \tag{5.19}$$

we find the determinant of the cofactor of either $\tilde{r}_{p+1}$-term must be zero. (The cofactor of a matrix element is the matrix remaining after deleting the row and column containing the element.) Each $\tilde{r}_{p+1}$ cofactor is the transpose of the other so they have the same determinant. When we set the determinant of the cofactor within the box in Eq. (5.17) equal to zero

$$\det \begin{vmatrix} r_1 & r_0 & r_1 & \cdots & r_{p-2} & r_{p-1} \\ r_2 & r_1 & \cdots & \cdots & r_{p-3} & r_{p-2} \\ \vdots & \vdots & \ddots & \vdots & \vdots & \vdots \\ \vdots & \vdots & \vdots & \ddots & \vdots & \vdots \\ r_p & r_{p-1} & \vdots & \vdots & r_1 & r_0 \\ \tilde{r}_{p+1} & r_p & r_{p-1} & \cdots & r_2 & r_1 \end{vmatrix} = 0 \tag{5.20}$$

we obtain an expression for the ME $\tilde{r}_{p+1}$ in the lower left corner. The ME estimate of the superresolved spectrum could then be obtained from Eq.

(5.11), with the sum over $n = -(p + 1)$ to $(p + 1)$ to include the $\tilde{r}_{-(p+1)}$ and $\tilde{r}_{p+1}$ terms. Note $\tilde{r}_{-(p+1)} = \tilde{r}_{p+1}$.

However, the spectrum $S(\omega)$ is usually expressed in terms of the parameters of an autoregressive (AR) all-pole filter, the properties of which will be described in the next section. In order to relate the ME results Eqs. (5.11) and (5.20) to the spectrum of this filter, we define a set of coefficients $\{a_i\}$ by the matrix of the first p rows and all $p + 1$ columns of (5.20) as

$$\begin{bmatrix} r_1 & r_0 & r_1 & \cdots & \cdots & r_{p-1} \\ r_2 & r_1 & r_0 & \cdots & r_{p-3} & r_{p-2} \\ \vdots & \vdots & \ddots & \vdots & \vdots & \vdots \\ \vdots & \vdots & \vdots & \ddots & \vdots & \vdots \\ r_p & r_{p-1} & r_{p-2} & \cdots & r_1 & r_0 \end{bmatrix} \begin{pmatrix} 1 \\ a_1 \\ \vdots \\ a_p \end{pmatrix} = (\mathbf{0}) \tag{5.21}$$

Equation (5.21) furnishes p equations for the $a_1, \ldots, a_p$. We shall show in the next section that these a_i-coefficients are the coefficients of the AR filters with the same $\{r_k\}$ and that Eq. (5.21) implies Eq. (520). The ME power spectum from the filter with white noise[2] input is

$$S(\omega)^{\mathrm{ME}} = \frac{\Delta t\, R_w}{\left| 1 + \sum_{k=1}^{p} a_k e^{-j\omega k \Delta t} \right|^2}, \quad \text{Burg ME Spectrum} \tag{5.22}$$

with R_w defined as

$$R_w = r_0 + a_1 r_1 + \cdots + a_p r_p = r_0 + \sum_{k=1}^{p} a_k r_k \tag{5.23}$$

Equation (5.22) is the famous Burg ME power spectrum [5.7], a different form of which was described by E. T. Jaynes as "one of the most beautiful analytical results in statistics" [5.5].

Strictly speaking, Eq. (5.22) is an ME spectrum for the AR filter only for Gaussian random processes and known uniformly spaced autocorrelation data $\{r_k\}$ [5.2, p. 199]. We regard the unbiased estimator Eq. (5.9) as a datum. If the $\{a_k\}$ are estimated as $\{\hat{a}_k\}$ (the hat, ˆ, denotes an estimated parameter) by the Burg-recursion or "harmonic" Burg method [see description just beyond Eq. (5.112)] then Eq. (5.22) is the "harmonic" Burg spectrum. Methods of estimating the $\{\hat{a}_k\}$ from data other than the known $\{r_k\}$ and/or nonuniformly spaced $\{r_k\}$ will yield an AR rather than a Burg ME spectrum Eq. (5.22).

[2] A white noise sequence is selected at random from a Gaussian probability distribution with zero mean and variance σ^2. A real-sequence autocorrelation function is σ^2 for zero lag, zero otherwise. Its power spectral density is σ^2. A complex-sequence $u_n + jv_n$ has u_n and v_n each real-white uncorrelated sequences with autocorrelation function $\sigma^2/2$ for zero lag. The power spectral density is again σ^2.

5.4 RELATIONSHIP OF THE GIBBS-BURG RESOLUTION TO AUTOREGRESSIVE (AR) ALL-POLE MODELING

We now show that the ME determinantal equation Eq. (5.20) for $\tilde{r}_{p+1}$ is implied by Eq. (5.21), in terms of the coefficients $\{a_i\}$ of a pth-order autoregressive all-pole filter. The output spectrum of this filter with white noise input is Eq. (5.22), where the numerator is identical to the input noise power. In effect, this filter output power spectrum models the spectrum of the ME ensemble of possible f-sequences, consistent with the measured $\{r_k\}$.

5.4.1 Extension of the Autocorrelation Data Sequence

An autoregressive process of order p is described by the equation

$$f_n + a_1 f_{n-1} + \cdots + a_p f_{n-p} = w_n \tag{5.24}$$

where the $\{a_k\}$ are filter coefficients and w_k is a white noise sample. According to this, the best estimated values, f_{ne}, of the f_n for any $n \geq p$ are

$$f_{ne} = -(a_1 f_{n-1} + \cdots + a_p f_{n-p}), \quad \text{any } n \geq p \tag{5.25}$$

Minimization of the mean-square prediction error [5.8]

$$\sum_{n=-\infty}^{\infty} (f_n - f_{ne})^2 \qquad \text{for deterministic signal } \{f_n\}$$

$$\langle (f_n - f_{ne})^2 \rangle \text{ for a stationary random signal,} \qquad \text{any } n \geq p$$

for variations in the $\{a_k\}$ leads to these equations for the filter $\{a_k\}$

$$\sum_{k=1}^{p} a_k r'_{n-k} = -r'_n, \quad \text{any } n$$

where

$$\begin{aligned} r'_j &= r_j, \text{ datum}, \quad |j| \leq p \\ &= \hat{r}_j, \text{ extrapolated}, \quad |j| > \text{p}. \quad r'_{-j} = r'_j \end{aligned} \tag{5.26}$$

Notice that these equations for $n = 1, 2, \ldots$, p reproduce Eq. (5.21). If we add to Eq. (5.26) the $j = 0$ equation

$$r_0 + a_1 r_{-1} + \cdots + a_p r_{-\text{p}} = \langle w_k^2 \rangle$$

the $0 \leq j \leq p$ equations are called the AR Yule-Walker normal equations [5.2, p. 183].

5.4.2 Equivalence of ME and AR Spectra

Now we are ready to prove that Eq. (5.26) for $1 \leq n \leq p$, which includes Eq. (5.21), implies the ME equations Eq. (5.20) for $\tilde{r}_{p+1}$. This will show that the spectrum Eq. (5.11) extended over $-\infty < n < \infty$ is identical to the Burg spectrum Eq. (5.22). To show that Eq. (5.20) is indeed true for the filter, we must show that the last row is a linear combination of the first p rows. Equation (5.26) for $j = p + 1$ says

$$\tilde{r}_{p+1} + a_1 r_p + a_2 r_{p-1} + \cdots + a_p r_1 = 0 \tag{5.27}$$

For $j = p$ Eq. (5.26) says

$$r_p + a_1 r_{p-1} + a_2 r_{p-2} + \cdots + a_p r_0 = 0 \tag{5.28}$$

and so on, for $j = p - 2, \ldots, 1$. The last equation for $j = 1$ is (with $r_{-j} = r_j$)

$$r_1 + a_1 r_0 + a_2 r_1 + \cdots + a_p r_{p-1} = 0 \tag{5.29}$$

This block of equations beginning with Eq. (5.27) says that $\tilde{r}_{p+1}$ in Eq. (5.20) is a linear combination of $r_p, \ldots, r_1$ above it; that r_p is the same linear combination of $r_{p-1}, \ldots, r_0$ above *it*; and so on, over to r_1 in the lower right hand corner of Eq. (5.20). Thus, the last row of Eq. (5.20) is a_1 times the preceding pth row plus a_2 times the $(p - 1)$ row plus . . . plus a_p times the first row. Therefore, the last row Eq. (5.20) is a linear combination of the preceding rows, the determinant is zero, and the ME $\tilde{r}_{p+1}$ is identical to $\tilde{r}_{p+1}$ of the pth order AR filter.

If we generate all the $\tilde{r}_{p+2}, \tilde{r}_{p+3}, \ldots$ by extrapolation according to Eq. (5.26) for the filter, the ME spectrum

$$S(\omega)^{\mathrm{ME}} = \Delta t \sum_{k=-\infty}^{\infty} r_k' e^{-j\omega k \Delta t}$$

will be the same as the filter spectrum. We proceed next to derive the latter.

5.4.3 Derivation of the AR Filter Spectrum

Let the z-transform of f_n be $F(z)$; of f_{n-i}, $z^{-i} F(z)$; so the z-transform of Eq. (5.24) is

$$F(z)[1 + a_1 z^{-1} + \cdots + a_p z^{-p}] = W(z) \tag{5.30}$$

$F(z)$ is the filter output so the filter transfer function $T(z)$ is

$$T(z) = \frac{F(z)}{W(z)} = [1 + a_1 z^{-1} + \cdots + a_p z^{-p}]^{-1} \tag{5.31}$$

The quantity [ ] indicates the filter has p finite poles.

The output spectrum of this filter is obtained by substituting $z = \exp(-j\omega\Delta t)$ into Eq. (5.30),

$$F(\omega) = \frac{W(\omega)}{1 + \sum_{k=1}^{p} a_k e^{-j\omega k\Delta t}} \tag{5.32}$$

and the output spectrum $S^F(\omega) = |F(\omega)|^2$ of the filter (F) is

$$S^F(\omega) = \frac{|W(\omega)|^2}{\left|1 + \sum_{k=1}^{p} a_k e^{-j\omega k\Delta t}\right|^2} \tag{5.33}$$

This is identical to Eq. (5.22) for the ME spectrum because $|W(\omega)|^2 = \Delta t R_w$ of Eq. (5.23). To see this, multiply Eq. (5.24) by w_n and sum over n. We obtain, by averaging,

$$\langle f_n w_n\rangle + a_1 \langle f_{n-1} w_n\rangle + \cdots + a_p\langle f_{n-p} w_n\rangle = \langle w_n^2\rangle \tag{5.34}$$

An expression for $\langle f_n w_n\rangle$ is obtained by multiplying Eq. (5.24) by f_n and averaging

$$r_0 + a_1 r_1 + \cdots + a_p r_p = \langle f_n w_n\rangle \tag{5.35}$$

Proceeding this way, we write Eq. (5.24) for w_{n-1}, multiply by w_n and average, then write Eq. (5.24) for w_{n-2}, multiply by w_n and average, and so on, to obtain a series of homogeneous equations for $\langle w_{n-1} w_n\rangle = 0$, $\langle w_{n-2} w_n\rangle = 0$, and so on. These show $\langle f_{n-1} w_n\rangle, \langle f_{n-2} w_n\rangle, \ldots, \langle f_{n-p} w_n\rangle$ are all zero—as expected by causality—in Eq. (5.34). Therefore, Eqs. (5.34) and (5.35) tell us

$$\langle w_k^2\rangle = r_0 + a_1 r_1 + \cdots + a_p r_p = R_w \text{ of Eq. (5.23)} \tag{5.36}$$

Equations (5.11) for the $n = 0$ term and (5.36) now tell us about the white noise spectrum

$$|W(\omega)|^2 = \Delta t\, (r_0)_{\text{noise}} = \Delta t\, \langle w_k^2\rangle = \Delta t\, R_w \tag{5.37}$$

which proves Eq. (5.22) for the ME spectrum is identical to the filter output spectrum, Eq. (5.33).

Summarizing, we see that the Gibbs-Burg ME method of superresolving the power spectrum by imbedding the signal in an ME ensemble of signals with the same average measured autocorrelations $r_0, \ldots, r_p$ and "extrapolated" $\tilde{r}_{p+1}$ is equivalent to computing the output spectrum of a pth-order AR all-pole filter driven by white noise. *The noise-driven filter physically realizes this ME spectrum.*

According to remarks following Eq. (5.23), the filter spectrum Eq. (5.33) would not strictly be equivalent to the ME spectrum Eq. (5.22) if the $\{a_k\}$ were not obtained from the autocorrelation data $\{r_k\}$ in Eq. (5.21).

The Gibbs-Burg spectrum is based on an ME probability distribution for the signal values $f_1, \ldots, f_{p+1}$ and extrapolated $\tilde{f}_{p+2}$ (a nonunique time series). In Sec. 5.5, we present a derivation of the same power spectrum from an ME Gibbs probability distribution for the spectral amplitudes S, defined in Eq. (5.4).

5.4.4 Limitations of the Gibbs-Burg Spectral Estimator, Eq. (5.22)

Both the ME and harmonic Burg (as well as the covariance and modified covariance) methods tend to have better frequency resolution than other AR algorithms for a given order p, primarily because of the absence of windowing [5.2, p. 225]. But the Burg spectral estimators Eq. (5.22) can perform poorly in some situations.

Resolution of sinusoids in white noise depends on the signal-to-noise ratio (SNR). As SNR decreases one or more of the sinusoids may be lost. Equation (5.22) for the harmonic Burg process might show spectral line splitting if (a) the SNR is high, (b) initial phases of sinusoidal components are odd multiples of 45°, (c) the raw data segment length allows sinusoidal components to be an odd number of quarter cycles long, or (d) the number, p, of AR parameters estimated is a large fraction of data samples used for the estimation.

The one-dimensional (time or space) ME analysis of Burg is not directly extendable to two and higher dimensions. Special AR techniques are employed to compute self- and cross-channel power spectral densities. For arbitrary arrays in two and three spatial dimensions the ME spectra are not related to the corresponding AR models as they are in one dimension [5.2, Chap. 16] [5.9].

5.5 SPECTRAL FREQUENCY RESOLUTION BY THE GIBBS METHOD (PROBABILITY DISTRIBUTION FOR THE POWER SPECTRUM)

We now follow Rietsch [5.10] and derive the Burg spectrum Eq. (5.22) from a Gibbsian probability distribution for the spectral power amplitudes S of Eq. (5.4) rather than from the distribution P of Eq. (5.15) for the temporal raw data amplitudes.[3] He also included a "weight factor" distribution $Q(\overline{S})$ which plays the role of a prior probability distribution r of Sec. 1.7.

The discretized autocorrelation function r_n is the discrete inverse Four-

[3] We replace Rietsch's parameters $\overline{P}$, $w(\overline{P})$, and A_n by our $\overline{S} = (S_{-M}, \ldots, S_M)$, $Q(\overline{S})$ and r_n in order to be consistent with the foregoing discussion. Rietsch's $(M - 1)/2$ is our M. The bar over a symbol denotes a vector.

ier transform of the power spectrum $[S_m\}$ from Eq. (5.4). By analogy with Eqs. (5.7) and (5.8) r_n is

$$r_n = r(n\Delta t) = \frac{\Delta\omega}{2\pi} \sum_{m=-M}^{M} S_m e^{+jm\Delta\omega n\Delta t}, \quad \Delta f = \frac{\Delta\omega}{2\pi} \tag{5.38}$$

With frequency discrimination $\Delta\omega = \omega_M/M$ and $\Delta t = \pi/\omega_M$, the exponent is $j\pi mn/M$.

Rietsch maximizes his Gibbsian entropy

$$h = -\int P(\overline{S}) \log\left[\frac{P(\overline{S})}{Q(\overline{S})}\right] d\overline{S} \tag{5.39}$$

$P(\overline{S})$ being the ensemble probability of spectrum $\overline{S}$, which is a limiting form of Shannon entropy for discrete variables $\{S_i\}$ as each $\Delta S_i \to 0$. This involves the multivariate probability distribution of the continuous $\{S_i\}$

$$P(\overline{S}) = P(S_{-M}, \ldots, S_0, \ldots, S_M) \tag{5.40}$$

subject to the constraints

$$\int P(\overline{S})\, d\overline{S} = 1 \tag{5.41}$$

and Eq. (5.38), with S_m replaced by the ensemble average $\langle S_m \rangle$

$$\Delta f \int P(\overline{S}) \left[\sum_m S_m e^{j\pi mn/M}\right] d\overline{S} = r_n, \qquad 0 \le n \le N \tag{5.42}$$

The procedure of Appendix A yields

$$P(\overline{S}) = Q(\overline{S}) \frac{\exp\left[-\sum_m L_m(\overline{\beta})\, S_m\right]}{Z(\overline{\beta})} \tag{5.43}$$

with partition function $Z(\overline{\beta})$ given by

$$Z(\overline{\beta}) = \int Q(\overline{S}) \exp\left[-\sum_m L_m(\overline{\beta})\, S_m\right] d\overline{S} \tag{5.44}$$

Here

$$L_m = \sum_{n=0}^{N} \beta_n e^{j\pi mn/M} \tag{5.45}$$

Rietsch then deduces from three facts that $Q(\overline{S})$ has the following form,

$$Q(\overline{S}) = \gamma^{2M+1} \exp[-\gamma \sum S_m] \tag{5.46}$$

The facts are that: (1) $S(\omega)$ is a sum of squares of Fourier sine and cosine components of ω, (2) these components are independent; that is, not constrained, and (3) their probability distributions do not depend on the origin of the time axis. When Eq. (5.46) is absorbed into Eqs. (5.44) and (5.45) and γ into each L_m there obtains

$$P(\bar{S}) = \prod_m P_m(S_m) \tag{5.47}$$

$$\langle S_m \rangle = \int S_m\, P_m(S_m)\, dS_m = \frac{1}{L_m} \tag{5.48}$$

If the (N + 1) unknown β_n were found self-consistently to satisfy the (N + 1) constraints Eq. (5.42), the ME discrete spectrum would be given by Eq. (5.48) over the bandwith $(-\omega_m, +\omega_M)$.

To obtain the Burg spectrum, we pass to the limit $T \to \infty$, $\Delta\omega \to 0$, still over the same bandwidth, and with the same discretization Δt, whereupon $m\Delta\omega \to \omega$ in Eq. (5.38) and

$$\begin{aligned} L_m \to L(\omega) &= \sum_n \beta_n e^{j\omega n \Delta t} \\ S_m \to S(\omega) &= \frac{1}{L(\omega)} \end{aligned} \tag{5.49}$$

With the aid of the Fejer-Riesz Theorem [5.11] $L(\omega)$ is factored into $|\,\Gamma G(\omega)\,|^2$ so that $S(\omega)$ can be written as

$$S(\omega) = \frac{1/\Gamma^2}{G(\omega)G^*(\omega)} = \frac{1/\Gamma^2}{\left| 1 + \sum_{\nu=1}^{N} g_\nu\, e^{j\omega\nu\Delta t} \right|^2} \tag{5.50}$$

This is nearly the Burg form, Eq. (5.22).

To convert Eq. (5.50) into Eq. (5.22) we multiply both sides by $G^*(\omega)$ $\exp(j\omega n\Delta t)$ and take $\int_{-\omega_M}^{\omega_M} d\omega$, obtaining

$$\sum_{\nu=0}^{N} g_\nu^* \int_{-\omega_M}^{\omega_M} d\omega\, S(\omega) e^{-j\omega(\nu - n)\Delta t} = \frac{1}{\Gamma^2} \int_{-\omega_M}^{\omega_M} d\omega \frac{e^{j\omega n\Delta t}}{\sum_{\nu=0}^{N} g_\nu\, e^{j\omega\nu\Delta t}} \tag{5.51}$$

From Eq. (5.38) in the limit of $\Delta\omega \to 0$ the left side of Eq. (5.51) becomes $\sum_{\nu=0}^{N} g_\nu^* r_{n-\nu}$. The right-side integral is the limit, as $y \to \infty$, of an integral in the complex ω plane over the closed contour, from $-\omega_M + j0$ to $+\omega_M$, up to $\omega_M + jy$, back to $-\omega_M + jy$, and down to $-\omega_M + j0$ [5.12]. It evaluates to $\delta_{n0}/(g_0\Delta t)$. Then, Eq. (5.51) reads

$$\sum_{\nu=0}^{N} g_\nu^* r_{n-\nu} = \frac{\delta_{n0}}{\Gamma^2\, g_0\, \Delta t}, \qquad g_0 = 1 \tag{5.52}$$

Equation (5.52) is identical to Eq. (5.26) for $n \neq 0$ if we identify $g_\nu^* = a_\nu$; for $n = 0$ Eq. (5.52) evaluates to R_w by Eq. (5.36). Therefore, $1/\Gamma^2 = R_w \, \Delta t$ and Eq. (5.50) becomes

$$S(\omega) = \frac{\Delta t \, R_w}{\left| 1 + \sum_{\nu=0}^{N} a_\nu \, e^{-j\omega\nu\Delta t} \right|^2} \qquad \text{(Burg ME)} \qquad (5.53)$$

the Burg ME spectrum of Eq. (5.22).

It is interesting to note that all three derivations of the Burg ME spectrum—by probability function for the time series amplitudes, AR all-pole filter, and probability function for the spectral amplitudes—are written for a *continuous* spectrum over $(-\omega_M, +\omega_M)$, as if the measurement interval T were infinite. Recall that the Gibbs-Burg derivation of probability $P(f_1, \ldots, \tilde{f}_{p+2})$ was based on extrapolation of the time series and autocorrelation sequence by one value each. This implies a gradually decreasing reliability in the ME prediction of $S(\omega)$ at successively smaller intervals of $\Delta\omega$, corresponding to lengthening effective observation times T′ beyond the measurement time T.

The Burg spectrum applies to a single measured time series at intervals $0, \Delta t, 2\Delta t, \ldots, T$. Suppose we return another day and obtain from the same process a different time series with a different Burg spectrum and different filter coefficients $\{a_k\}$! This could be due to Statistical Uncertainty as described in Sec. 3.8. More likely it is caused by noise in the time series or in the measuring instruments. Or it could indicate the fact that the *process itself* has changed, which means it is not stationary in time. Only a supplemental analysis can resolve this matter.

5.6 SPATIAL ARRAY DATA PROCESSING (SPACE-WAVENUMBER DOMAINS AT A FIXED FREQUENCY)

Electromagnetic antenna arrays are employed to process waves coming from an object in space and resolve its direction and perhaps velocity. Both electromagnetic and seismic arrays are used to locate sources of underground explosions and earthquakes. Spatially linear arrays of uniformly spaced elements are usually employed for direction finding because the data processing is simplified. In this section, we make the usual assumption that the array is so far away from the source that any arriving wave has plane wave form, the field is uniform at a given time at any position in a plane perpendicular to the direction of travel of the wave.

A plane wave at circular frequency $\omega = 2\pi f$, traveling in the $x - z$ plane, is characterized by a space-time factor

$$\exp[j(\omega t - \bar{k}\cdot\bar{r})] = \exp[j(\omega t - k_x x - k_y y - k_z z)], \qquad k_y = 0 \qquad (5.54)$$

k_x and k_z are the wavenumbers[4] in the x- and z-directions, identified in Fig. 5.1. We observe from the definition of angle θ_k that

$$k_x = k \sin \theta_k, \qquad k_z = k \cos \theta_k \tag{5.55}$$

The phase velocity v of the wave in its direction of travel $\bar{k}$ is deduced by writing $(\omega t - kr)$ as $[t - r/(\omega/k)] = (t - r/v)$, whereupon

$$v = \frac{\omega}{k} \tag{5.56}$$

The phase velocity in the x-direction is found by holding z fixed, say at $z = 0$, and rewriting $(\omega t - k_x x)$ as $(t - x/v_x)$. Thus,

$$v_x = \frac{\omega}{k_x} = \frac{\omega}{k \sin \theta_k} = \frac{v}{\sin \theta_k}, \qquad | v_x | \geq v \tag{5.57}$$

Analogously, the phase velocity v_z in the z-direction is

$$v_z = \frac{\omega}{k_z} = \frac{\omega}{k \cos \theta_k} = \frac{v}{\cos \theta_k}, \qquad | v_z | \geq v \tag{5.58}$$

The wavelengths are related to the velocities and frequency $f = \omega/2\pi$ in the usual way, namely,

$$v = f \lambda, \qquad \lambda = \frac{2\pi}{k} \tag{5.59}$$

$$| v_x | = f \lambda_x, \qquad \lambda_x = \frac{2\pi}{| k_x |} = \frac{\lambda}{| \sin \theta_k |} \tag{5.60}$$

$$| v_z | = f \lambda_z, \qquad \lambda_z = \frac{2\pi}{| k_z |} = \frac{\lambda}{| \cos \theta_k |} \tag{5.61}$$

Equation (5.60) says that v_x increases as θ_k approaches 0 or π and the wave propagates more nearly vertically. This is because wavelength λ_x, the distance the wave travels along x in one period $(1/f)$, lengthens as θ_k decreases. Similarly, when $| \theta_k |$ increases toward $\pi/2$ and the wave travels more nearly horizontally both $| v_z |$ and λ_z increase.

We now begin a discussion of array autocorrelation data processing to ascertain the directions of various incoming waves with k-vectors in the $x - z$ plane. The x axis is oriented along the array, as shown in Fig. 5.2. The array elements are spaced at equal intervals Δx, analogous to the sampling time interval Δt of time-frequency data analysis. The total spatial extent of the array is L_x, corresponding to total duration T of a temporal signal. The advantage of using an array of equispaced elements is that we can super-

[4] Often $1/\lambda$, inverse wavelength, is defined as wavenumber instead.

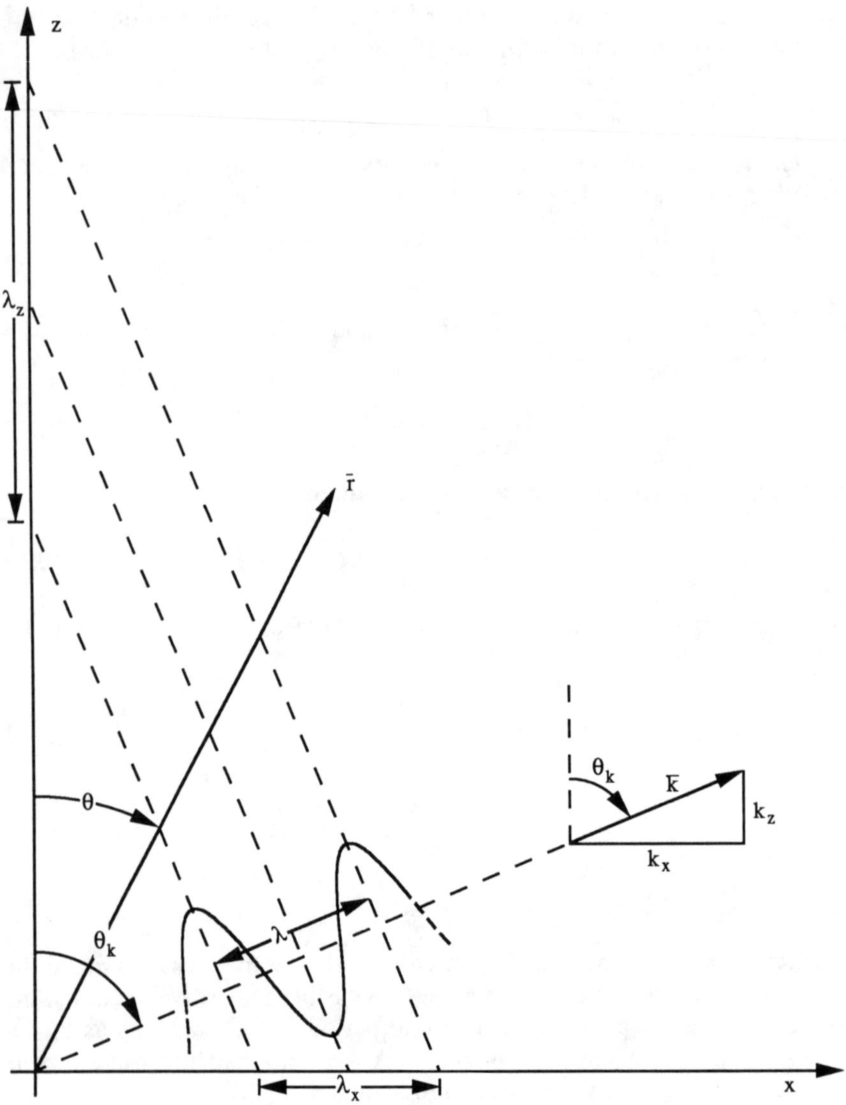

Figure 5.1 A portion of a plane wave of frequency ω, one wavelength long between consecutive peaks, traveling in the x − z plane in direction $\bar{k}$ with velocity $v = \omega/k$.

resolve the k_x spectrum of the waves by the Gibbsian ME method of Sec. 5.3 by analogy with time-frequency data processing.

A coherent component in a plane wave traveling in direction $\bar{k}$ is represented as $A_i \exp[j(\omega t - \bar{k}\cdot\bar{r}_i)]$ at frequency ω and position $\bar{r}_i$ in the array. An incoherent component would have a random phase $\phi_i(t)$ in place of ωt.

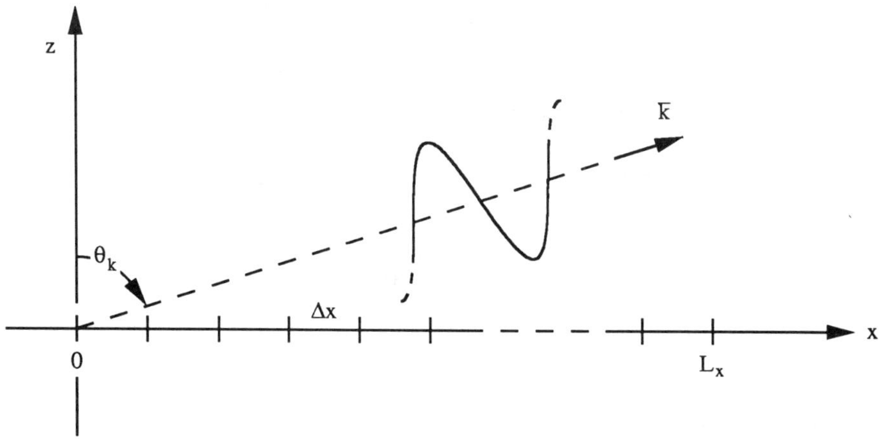

Figure 5.2 A spatial array along the x-axis to resolve the k_x wavenumber spectrum of waves traveling in $x - z$ plane. Correlation data processing is done at a chosen circular frequency ω_o.

The Fourier transform pair Eqs. (5.1) and (5.2) is replaced by the two-dimensional expressions

$$\frac{\mathrm{F}(d\omega, dk_x)}{d\omega\, dk_x} = \int \mathrm{dx} \int dt\, f(x, t)e^{-j(\omega t + k_x x)} \tag{5.62}$$

$$f(x, t) = \frac{1}{(2\pi)^2} \int \int \mathrm{F}(d\omega, dk_x)e^{+j(\omega t + k_x x)} \tag{5.63}$$

for the spatial array. Here, $\mathrm{F}(d\omega, dk_x)$ denotes a partially or fully incoherent amplitude spectrum in a narrow band $d\omega dk_x$ at ω, k_x. The autocorrelation function for waves that are partially or fully temporally incoherent, analogous to $r(\tau)$ of Eq. (5.3), is now generally a complex function of spatial autocorrelation distance ρ_x as well as temporal lag τ,

$$\begin{aligned} r(\rho_x, \tau) &= \langle < f^*(x, t) f(x + \rho_x, t + \tau) >_x \rangle_t \\ &= \frac{1}{\mathrm{L}_x \mathrm{T}} \int_0^{\mathrm{L}_x} dx \int_0^{\mathrm{T}} dt\, f^*(x, t)\, f(x + \rho_x, t + \tau) \end{aligned} \tag{5.64}$$

The $\ll \gg$ denotes averaging over both x and t, the latter performed over snapshots of the array element fields at different times. If we replace f^* and f here with their spectral expansions Eq. (5.63) we can write Eq. (5.64) in the form

$$r(\rho_x, \tau) = \int \int \langle < | \mathrm{Z}(d\omega, dk_x) |^2 > \rangle e^{j(\omega\tau + k_x \rho_x)} \tag{5.65}$$

where $Z(d\omega, dk_x) = F(d\omega, dk_x)/(2\pi)^2$ is a convenient representation to relate the measured $r(\rho_x, \tau)$ to the wavenumber spectrum $S(k_x, \omega)$.

It is convenient to suppress the time dimension in spatial array processing to resolve the wavenumber spectrum with one-dimensional spatial autocorrelation data. This is done by filtering the raw data $f(x, t)$ over time T within a narrow band $\Delta\omega$ at an appropriate frequency ω_0. Most electromagnetic and seismic records will allow this because they are long compared to the period $2\pi/\omega_0$. Writing $Z(d\omega, dk_x) = Z(\Delta\omega, dk_x)$ for this band and removing $\exp(j\omega_0\tau)$ to the left side of Eq. (5.65) we have

$$r_{\omega_0}(\rho_x, \tau)e^{-j\omega_0\tau} = \int \langle < | Z_{\omega_0}(\Delta\omega, dk_x) |^2 > \rangle e^{jk_x\rho_x} \tag{5.66}$$

The wavenumber spectrum $S(\omega_0, k_x)$, written more succinctly as $S_{\omega_0}(k_x)$, is related to $r(\rho_x, \tau)$ as

$$r(\rho_x, \tau) = \frac{1}{(2\pi)^2} \int dk_x \int d\omega\, S_\omega(k_x) e^{j(k_x\rho_x + \omega\tau)} \tag{5.67a}$$

and in a narrow band around ω_0 it is

$$r_{\omega_0}(\rho_x, \tau)e^{-j\omega_0\tau} = \frac{1}{(2\pi)^2} \int dk_x\, \Delta\omega\, S_{\omega_0}(k_x) e^{jk_x\rho_x} \tag{5.67b}$$

We see from Eqs. (5.66) and (5.67b) that

$$\langle < | Z_{\omega_0}(\Delta\omega, dk_x) |^2 >_x \rangle_t = \frac{1}{(2\pi)^2} \Delta\omega\, dk_x S_{\omega_0}(k_x) \tag{5.68}$$

We want to write the spectrum $S_{\omega_0}(k_x)$ in terms of $r_{\omega_0}(\rho_x, \tau)$ for Gibbs-Burg autocorrelation data processing. Inversion of Eq. (5.67) gives at ω_0,

$$S_{\omega_0}(k_x) = \int d\rho_x \left[\int d\tau\, r_{\omega_0}(\rho_x, \tau) e^{-j\omega_0\tau} \right] e^{-jk_x\rho_x} \tag{5.69}$$

This can be verified by inverting Eq. (5.67b) to obtain $\Delta\omega S_{\omega_0}(k_x)/2\pi$, then taking $\int d\tau$ over $T = 2\pi/\Delta\omega$.

The quantity within [ ] in Eq. (5.69) is called the *power spectrum* of the spatial array data processing and is denoted by a script $\mathcal{P}_{\omega_0}(\rho_x)$,

$$\mathcal{P}_{\omega_0}(\rho_x) = \int d\tau\, r_{\omega_0}(\rho_x, \tau) e^{-j\omega_0\tau} \quad \text{(watt)} \tag{5.70}$$

In terms of this, Eq. (5.69) is

$$S_{\omega_0}(k_x) = \int d\rho_x\, \mathcal{P}_{\omega_0}(\rho_x) e^{-jk_x\rho_x} \tag{5.71}$$

This is the relation for Gibbs-Burg data processing of $\mathcal{P}_{\omega_0}(\rho_x)$ to obtain wavenumber spectrum $S_{\omega_0}(k_x)$, analogous to the formula for the power frequency spectrum $S(\omega)$ of Eq. (5.4) and S_{AUTO} of Eq. (5.11).

TABLE 5.1 Correspondences between the Parameters for Time-Frequency Data Processing and Space-Wavenumber Processing

Time-Frequency, $f(t) \leftrightarrow F(\omega)$		Space-Wavenumber, $f(x) \leftrightarrow F(k_x)$
$t, \Delta t$		$x, \Delta x$
$\omega, \Delta\omega$		$k_x, \Delta k_x$
$\Delta t = \pi/\omega_M$	sampling	$\Delta x = \pi/(k_x)_M$
$\Delta\omega = 2\pi/T$	relations	$\Delta k_x = 2\pi/L_x$
(Frequency f (Hertz)		inverse wavelength, $1/\lambda$ (length^{-1})

The Gibbs data inversion of Sec. 5.5 would utilize the discretized Fourier mate of Eq. (5.71) for constraint information in deriving the Burg spectrum for $S_{\omega 0}(k_x)$.

Table 5.1 shows the correspondences between the earlier data processing in the time-frequency domains and in the space-wavenumber domains. We observe, in comparing the earlier equations with their analogs for the time-frequency processing, that there is one additional step in spatial array processing, namely computation of the power spectrum $\mathcal{P}_\omega$ as the Fourier transform of the double functional autocorrelation coefficient $r(\rho_x, \tau)$ by Eq. (5.70).

Now, if the plane waves impinging on the spatial array have wavelengths with a lowerbound λ then $k = 2\pi/\lambda$ has an upperbound, as does k_x. Therefore, $f(x, t)$ is bandlimited in k, as is $\mathcal{P}_\omega(\rho_x)$ in ρ_x. $\mathcal{P}_\omega$ is specified exactly by the signal samples $f(x)$ at intervals $n\Delta x$, Δx being the limit of spatial discrimination as related to $(k_x)_M$ in Table 5.1. The sampling criteria for the space-wavenumber domains in the table imply the same number of wavenumber intervals resolved over $[-(k_x)_M, +(k_x)_M]$ as spatial intervals over $(0, L_x)$

$$\frac{2(k_x)_M}{\Delta k_x} = \frac{L_x}{\Delta x} = \frac{(k_x)_M L_x}{\pi} \tag{5.72}$$

which is analogous to Eq. (5.6).

We now summarize the steps in the sampled-data processing of wavenumber spectrum $S_{\omega 0}(k_x)$ according to Eqs. (5.64), (5.70), and (5.71). There are $M + 1$ k_x points, $N + 1$ x points, and $Q + 1$ t points, at intervals of Δk_x, Δx, and $\Delta t = \pi/\omega_0$, respectively. First we evaluate r_{nq} $(n\Delta x, q\Delta t)$ from Eq. (5.64):

$$r_{nq}(n\Delta x, q\Delta t) = \frac{1}{NQ} \sum_{n'} \sum_{q'} f^*_{n'q'} f_{n'+n, q'+q} \tag{5.73}$$

This determines $\mathcal{P}_{\omega 0}(\rho_x = n\Delta x)$ from Eq. (5.70) as

$$\mathcal{P}_{\omega 0}(n\Delta x) = \Delta t \sum_q r_{nq} e^{-j\omega_0 q\Delta t} \tag{5.74}$$

By defining complex amplitudes F_{nq} by

$$f_{nq} = F_{nq}e^{j\omega_0 q\Delta t} + F^*_{nq}e^{-j\omega_0 q\Delta t} \tag{5.75}$$

where F_{nq} has a fixed or random phase if coherent or incoherent, respectively, [see definition preceding Eq. (5.62)] we can simplify Eq. (5.74) to

$$\mathcal{P}_n = \Delta t \sum_q R_{nq} = \mathcal{P}^*_{-n} \tag{5.76}$$

where

$$R_{nq} = \frac{1}{NQ}\sum_{n'q'} F^*_{n'q'}F_{n'+n,q'+q} \tag{5.77}$$

This has a value if the wave is coherent or incoherent, provided the direction of the spatial wave is such that the net phase of $F^*_{n'q'}F_{n'+n,q'+q}$ is zero at the two space-time points (n', q') and (n' + n, q' + q).

Equation (5.71) yields the discrete wavenumber spectrum $S_{\omega_0}(k_x = m\Delta k_x)$ as

$$S_{\omega_0}(m\Delta k_x) = \Delta x \sum_n \mathcal{P}_n e^{-jm\Delta k_x n\Delta x} \tag{5.78}$$

The Gibbs-Burg wavenumber spectrum superresolution is based on Eq. (5.78) and proceeds by direct analogy with the superresolution of a frequency spectrum according to Eq. (5.11). But, in order to have Eq. (5.78) analogous to Eq. (5.11) and Eq. (5.76) analogous to Eq. (5.9), we must show $\mathcal{P}_n$ has an autocorrelation form. This is apparent if we substitute Eq. (5.77) into Eq. (5.76), note $\Delta t = T/Q$, and define new complex amplitudes

$$\mathcal{F}_{n'+n} = \sqrt{\frac{T}{N}}\frac{1}{Q}\sum_q F_{n'+n,q'+q} \tag{5.79a}$$

independent of q′ because $\omega_0 q\Delta t$ is periodic over Q, and

$$\mathcal{F}_{n'} = \sqrt{\frac{T}{N}}\frac{1}{Q}\sum_{q'} F_{n'q'} \tag{5.79b}$$

then, $\mathcal{P}_n$ has the form

$$\mathcal{P}_n = \sum_{n'} \mathcal{F}^*_{n'}\mathcal{F}_{n'+n} \tag{5.80}$$

By analogy the ME complex joint probability density P ($\mathcal{F}_1, \mathcal{F}^*_1, \ldots, \mathcal{F}^*_{N+1}, \tilde{\mathcal{F}}_{N+2}, \tilde{\mathcal{F}}^*_{N+2}$ analogous to Eq. (5.13) will maximize the Shannon entropy, and the complex autocorrelation matrix replacing Eq. (5.17) is

$$\mathcal{R}_{N+1} \stackrel{\text{def}}{=} \begin{pmatrix} \mathcal{P}_0 & \mathcal{P}_1^* & \mathcal{P}_2^* & \cdots & \cdots & \mathcal{P}_N^* & \tilde{\mathcal{P}}_{N+1}^* \\ \mathcal{P}_1 & \mathcal{P}_0 & \cdots & \cdots & \cdots & \mathcal{P}_{N-1}^* & \mathcal{P}_N \\ \mathcal{P}_2 & \mathcal{P}_1 & \mathcal{P}_0 & \cdots & \cdots & \vdots & \vdots \\ \vdots & \vdots & \vdots & & \vdots & \vdots & \vdots \\ \vdots & \vdots & \vdots & \vdots & & \vdots & \vdots \\ \mathcal{P}_n & \mathcal{P}_{N-1} & \cdots & \cdots & \cdots & \mathcal{P}_0 & \mathcal{P}_1^* \\ \tilde{\mathcal{P}}_{N+1} & \mathcal{P}_N & \cdots & \cdots & \mathcal{P}_2 & \mathcal{P}_1 & \mathcal{P}_0 \end{pmatrix} \tag{5.81}$$

Matrix Eq. (5.17) is replaced by the new one within the box of Eq. (5.81). The resultant ME spectrum is

$$S_{\omega 0}(k_x)^{ME} = \frac{\Delta x \mathcal{P}_w}{\left| 1 + \sum_{n=1}^{N} a_n e^{-jk_x n \Delta x} \right|^2}, \quad \text{Burg ME form} \tag{5.82}$$

with

$$\mathcal{P}_w = \mathcal{P}_0 + \sum_{n=1}^{N} a_n \mathcal{P}_n^* \tag{5.83}$$

analogous to Eq. (5.23), and $\Delta x \mathcal{P}_w$ is the white noise spectrum feeding the Nth order AR filter to produce the output spectrum $S_{\omega 0}(k_x)^{ME}$ according to remarks analogous to those preceding Eq. (5.37).

5.7 TIME SERIES ANALYSIS BY AUTOREGRESSIVE MOVING AVERAGE (ARMA) AND MOVING AVERAGE (MA) METHODS

These alternate methods of waveform analysis are discussed briefly to put them in perspective with the autoregressive (AR) all-pole method, of which the Burg ME and harmonic techniques are special cases.

The ARMA time series model of order (p, q) is an extension of Eq. (5.24) to include earlier white noise driving terms on the right side,

$$f_n + a_1 f_{n-1} + \cdots + a_p f_{n-p} = b_0 w_n + b_1 w_{n-1} + \cdots + b_q w_{n-q} \tag{5.84}$$

$b_0 = 1$ is usually assumed and implies scaling the input w_n. If all the $\{a_i\}$ on the left are zero we have an MA model of raw data generation,

$$f_n = w_n + b_1 w_{n-1} + \cdots + b_q w_{n-q} \tag{5.85}$$

The power spectral density of ARMA (p, q) is [5.2, p. 175]

$$S_{ARMA}(f) = \Delta t <w_k^2> \left| \frac{B(f)}{A(f)} \right|^2 \tag{5.86}$$

where $\langle w_k^2 \rangle$ is the autocorrelation function r_0 of white noise and

$$A(f) = 1 + \sum_{k=1}^{p} a_k e^{-j2\pi fk\Delta t} \tag{5.87}$$

$$B(f) = 1 + \sum_{k=1}^{p} b_k e^{-j2\pi fk\Delta t} \tag{5.88}$$

Any additional observation noise present must be modeled separately on the right side of Eq. (5.84).

The power spectral density of the MA process is

$$S_{MA}(f) = \Delta t \, \langle w_k^2 \rangle \, | \, B(f) \, |^2 \tag{5.89}$$

Estimates of ARMA parameters $\{a_k\}$ and $\{b_k\}$ or MA parameters $\{b_h\}$ require the solution of nonlinear equations, which is one reason why AR data processing is more popular.

Despite the generality of the ARMA model few algorithms are available to estimate $S_{ARMA}(f)$ of Eq. (5.86). This is because of the difficulty of solving self consistently for the AR $\{a_k\}$ and MA $\{b_k\}$. Techniques have been developed based on a least-squares criterion and solution of linear equations to estimate first the AR- and then MA-parameters. A method is available for simultaneously estimating the AR- and MA-parameters by a linear method and then improving the estimates by iteration. Another method involves a time-sequential algorithm for simultaneous estimation of both parameter sets. Selection of the model orders p and q is usually ambiguous.

The MA power spectral density is appropriate for broad peaks and deep nulls but not for narrow peaks. There is a high p-order AR approximation to the MA process that employs only linear operations. Selection of model order q is ambiguous.

Fortunately an ARMA or MA process can be represented closely by an AR model of sufficiently high order. So one sometimes solves such a model in the first step of estimating the parameters of an ARMA or MA model.

5.8 THE PRONY METHOD OF AUTOCORRELATION AND RAW DATA MODELING

This method is more appropriate for power spectral density estimation if the data are known to consist solely of sinusoids in noise.

Many physical processes can be represented by an ARMA, AR, or MA model with white noise input and temporal output in which the spectral peaks characterize the process. An example is the electroencephalogram (EEG), with spectral peaks whose parameters are directly related to neurophysiological quantities (δ-waves, 0-4 Hz; θ-waves, 4-8 Hz; α-waves, 8-13 Hz; and β-waves above 13 Hz. These frequency ranges are approximate). Prony's method represents raw-sampled data $\{f_n = f(n\Delta t)\}$, autocorrelation data $\{r_n\}$, or the impulse response $h(t_n)$ of a causal spectral shaping system by a sum of exponentials,

$$f_n = \sum_{k=1}^{p} A_k e^{s_k n\Delta t} \tag{5.90}$$

$$\hat{r}_n = \hat{r}(\tau_n) = \sum_{k=1}^{p} \hat{R}_k e^{s_k|\tau_n|}, \qquad \tau_n = n\Delta t \tag{5.91}$$

$$h(t) = \sum_{k} B_k e^{s_k t} \tag{5.92}$$

with complex poles $s_k = -\alpha_k + j\omega_k$ and residues A_k, $\hat{R}_k$, and B_k. The hat ˆ denotes an estimated quantity.

We first describe the Prony estimation of the arrays $\{s_k\}$ and $\{\hat{R}_k\}$ from estimated $\{\hat{r}_k\}$ using the ARMA (p, q) model of Eq. (5.84). For the representation Eq. (5.92) to be valid, $q = p - 1$. The ARMA recursive equation is

$$f_k + \hat{a}_1 f_{k-1} + \cdots + \hat{a}_p f_{k-p} = w_k + b_1 w_{k-1} + \cdots + b_{p-1} w_{k-(p-1)} \tag{5.93}$$

Subsequently, we will apply the Prony method to estimation of $\{s_k\}$ and $\{A_k\}$ from the raw-data $\{f_k\}$ using the ARMA $(p, 0)$ or AR model.

Both methods have been successfully applied to interpret simulated EEG data with estimated pole order $\hat{p}$ in the range of 7 to 13 [5.13]. A patient in surgery has a temporally changing EEG (nonstationary), and for spectral estimation it was found that a raw-data segment length of 20 to 30 seconds appeared to offer a reasonable compromise between high stationarity (short segments with high ensemble variability) on one hand and low variability with good spectral resolution (long segments, *if* they were stationary) on the other. Experience indicates that the AR-Prony approach is simpler and more precise and more promising for surgical use. The amplitudes of the spectral peaks in an EEG appear to be a better indicator of the patient's health than do the locations of the $\{s_k\}$ [5.14].

The so-called square or determined Prony method is employed to solve for the poles $\{s_k\}$ with the estimated autocorrelations fitted exactly according to Eq. (5.91) as

$$\hat{r}_n = \sum_{k=1}^{p} \hat{R}_k z_k^n, \qquad z_k = e^{s_k\Delta t}, \qquad 0 \le n \le 2p - 1 \tag{5.94}$$

The procedure described by Marple [5.2, Sec. 11.4] for raw data $f_1, \ldots,$

$f_p, f_{p+1}, \ldots, f_{2p}$, revised for estimates $\hat{r}_0, \ldots, \hat{r}_{p-1}, \hat{r}_p, \ldots, \hat{r}_{2p-1}$ yields

$$\sum_{k=1}^{p} \hat{a}_k \hat{r}_{j-k} = -\hat{r}_j, \qquad j = p, \ldots, 2p - 1 \tag{5.95}$$

Equation (5.95) is a subset of the Yule-Walker equations [5.8]. The $\{\hat{a}_k\}$ are the coefficients of a polynomial of degree p that has the $\{z_k\}$ as its roots,

$$z^p + \hat{a}_1 z^{p-1} + \cdots + \hat{a}_p = 0 \tag{5.96}$$

The set of p equations Eq. (5.95) is first solved for the $\{\hat{a}_k\}$, then the $\{z_k\}$ are found from Eq. (5.96) and the poles $\{s_k\}$ by Eq. (5.94).

If $\mathrm{L} > 2p - 1$ estimates $\hat{r}_j$ are available, the Overdetermined Prony method yields estimates $\{s_k\}$ and $\{\hat{\mathrm{R}}_k\}$ for Eq. (5.94) to fit the waveform $\{\hat{r}_j\}$ in a least-squares sense [5.13],

$$\min_{\{\hat{\mathrm{R}}_i\}} \sum_{k=0}^{\mathrm{L}} \left(\sum_{i=1}^{p} \hat{\mathrm{R}}_i e^{s_i k \Delta t} - \hat{r}_k \right)^2 \tag{5.97}$$

The result of this, after rearrangement, is the set of p equations

$$\sum_{k=1}^{p} \hat{a}_k \left(\sum_{m=p}^{\mathrm{L}} \hat{r}_{m-j} \hat{r}_{m-k} \right) = - \sum_{m=p}^{\mathrm{L}} \hat{r}_{m-j} \hat{r}_m, \qquad 1 \le j \le p \tag{5.98}$$

to be solved for the $\{a_k\}$. These are inserted into Eq. (5.96), which is solved for the $\{z_k\}$ and then the $\{s_k\}$.

The parenthetical sum in Eq. (5.98) is the finite-measurement approximation to the covariance element $\langle r_j r_k \rangle$, the $\langle \quad \rangle$ denoting ensemble average.

This Overdetermined Prony method is equivalent to Makhoul's covariance method [5.8] applied to estimates $\hat{r}_k$ for $0 \le k \le 2p - 1$. That method computes each prediction error or residual, e_k, as

$$e_k = \hat{r}_k - \hat{r}_{ke} = \hat{r}_k - (-\hat{a}_1 \hat{r}_{k-1} - \cdots - \hat{a}_p \hat{r}_{k-p}) \tag{5.99}$$

[See Eq. (5.95)] over the points $k = p, p + 1, \ldots, \mathrm{L} > (2p - 1)$. Then, the least-squares minimization of

$$\sum_{k=p}^{\mathrm{L}} e_k^2 \tag{5.100}$$

relative to changes in $\{\hat{a}_k\}$ yields Eq. (5.98) again.

If $\mathrm{L} = 2\mathrm{p} - 1$ in Eq. (5.98) the set of p equations is equivalent to the determined set Eq. (5.95) because the matrix

$$\begin{bmatrix} \hat{r}_{L-p} & \cdots & \cdots & \hat{r}_{L-1} \\ \vdots & \ddots & \vdots & \vdots \\ \vdots & \vdots & \ddots & \vdots \\ \hat{r}_0 & \cdots & \cdots & \hat{r}_{p-1} \end{bmatrix}, \quad L = 2p - 1 \tag{5.101}$$

can be factored out of Eq. (5.98) written in matrix form.

Both the Square and Overdetermined Prony methods can be applied to the raw data Eq. (5.90) instead of autocorrelation estimates Eq. (5.91) with minor notational changes, as follows. Instead of autocorrelation estimates $\hat{r}_0, \ldots, \hat{r}_{p-1}, \ldots, \hat{r}_L$ ($L \geq 2p - 1$) we have raw (no hat) data $f_1, \ldots, f_p, \ldots, f_N$ ($N \geq 2p$). Equation (5.90) is

$$f_n = \sum_{k=1}^{p} A_k z_k^{n-1}, \qquad 1 \leq n \leq 2p \tag{5.102}$$

The Square Prony method for $N = 2p$ [5.2, p. 308] now yields

$$\sum_{k=1}^{p} a_k f_{j-k} = -f_j, \qquad j = p + 1, \ldots, 2p \tag{5.103}$$

where the $\{a_k\}$ are the coefficients of Eq. (5.96), solved for the roots $\{z_k\}$ and the poles $\{s_k\}$ in Eq. (5.94).

If $N > 2p$ raw data are available, the Overdetermined method yields equations for the $\{a_k\}$ analogous to Eq. (5.98),

$$\sum_{k=1}^{p} a_k \left(\sum_{m=p+1}^{N} f_{m-j} f_{m-k} \right) = - \sum_{m=p+1}^{N} f_{m-j} f_m, \qquad j = 1, \ldots, p \tag{5.104}$$

These are solved for the $\{a_k\}$, which are inserted into Eq. (5.96) to yield the $\{z_k\}$ and $\{s_k\}$.

The Overdetermined Prony method applied to raw data yields the same result as Makhoul's covariance method applied to the AR of Eq. (5.93) with all $b_k = 0$. The kth prediction error is written analogous to Eq. (5.99), as

$$e_k = f_k - f_{ke} = f_k - (-a_1 f_{k-1} - \cdots - a_p f_{k-p}) \tag{5.105}$$

over the points $k = p + 1, \ldots, N$. Then, the least-squares minimization of Eq. (5.100) yields Eq. (5.104) again.

If $N = 2p$ Eq. (5.104) is the same as Eq. (5.103) when the matrix

$$\begin{bmatrix} f_{N-p} & \cdots & \cdots & f_{N-1} \\ \vdots & \ddots & \vdots & \vdots \\ \vdots & \vdots & \ddots & \vdots \\ f_1 & \cdots & \cdots & f_p \end{bmatrix}, \quad N = 2p \tag{5.106}$$

is factored out of Eq. (5.104).

5.9 NOISE AND SUPERRESOLUTION

5.9.1 Two Coherent Sinusoidal Signals

Marple [5.15] has investigated the effect of noise and signal-to-noise ratio (SNR) on the ME resolution of two coherent equiamplitude sinusoidal signals very close in frequency. He has shown how one may obtain superresolution better than the Nyquist-criteria limit even for SNR ratios considerably less than 1 (< 0 dB). His analysis is summarized and his superresolution curves are reproduced as a function of both SNR and M, the order of the equivalent autoregressive all-pole filter. [$M \equiv p$ in Eq. (5.24).]

Marple considered two complex sinusoids of frequencies f_1 and f_2, carrying power P_s each, contaminated by white noise of unit power spectral density. The SNR is $10 \log(P_s)$ for each sinusoid. The spectral power response S was computed, either numerically or analytically, as a function of frequency, and the sinusoids were considered just resolved if $S(f_c)$ at the center frequency $f_c = (f_1 + f_2)/2$ was just equal to the average of the two spectra at f_1 and f_2

$$S(f_c) = \frac{1}{2}[S(f_1) + S(f_2)] \tag{5.107}$$

If f_1 were any closer to f_2 the spectral peaks were considered merged and the individual frequencies unresolved.

The frequency interval $(-f_M, +f_M)$ was divided into M intervals so $f_M = M\Delta f/2$. M intervals of time were also selected for the sampled-waveform analysis, so the maximum time[5] $T = M\Delta t$. A resolution parameter was defined as

$$R = 2\pi M \Delta t \Delta f = 2\pi T \Delta f \tag{5.108}$$

According to the sampling criteria $M\Delta t = T = 1/\Delta f$, so one would expect $R = 2\pi$. This is plotted as the "Uncertainty Principle" line on Fig. 5.3. Any resolution smaller than Δf would lower R and be called *superresolution* in the figure.

The Burg ME power spectrum Eq. (5.22) for an M-order AR filter, with the $\{a_k\}$ determined by the known autocorrelation function,

$$S(\omega = 2\pi f) = \frac{S_0}{\left| 1 + \sum_{k=1}^{M} a_k e^{-j\omega k \Delta t} \right|^2} \tag{5.109}$$

was then used to evaluate each term of the resolution equation Eq. (5.107),

[5] Marple's ΔT is our T.

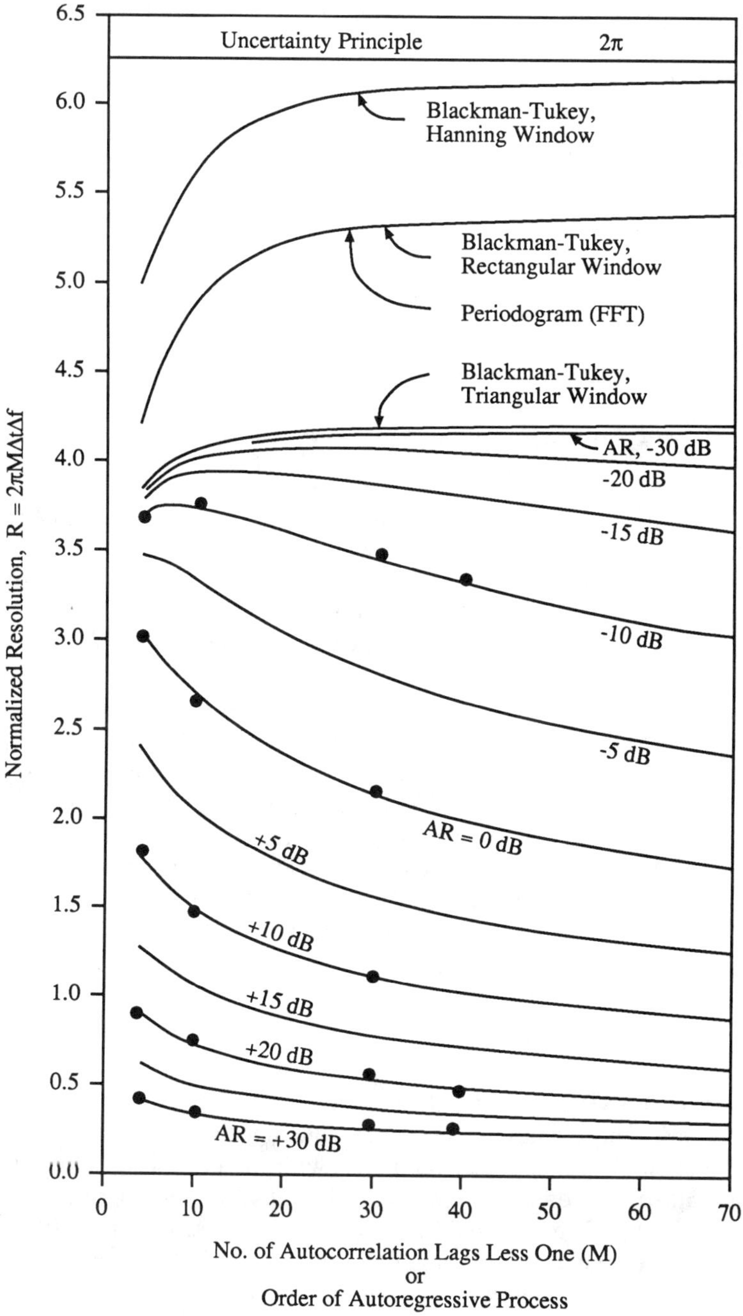

Figure 5.3 Resolution of two coherent sinusoidal frequencies with additive white noise, versus order of autoregressive process, with SNR as a parameter. Results for three Blackman-Tukey windows, one a Periodogram (FFT), are compared to the ME AR resolutions. Black dots indicate mean resolution of an ensemble of computer simulations. From [5.15] Fig 2.

and this yielded the following resolution condition, quoted from [5.15]

$$| 1 + (M + 1)P_s - P_s C_M(\Delta\gamma) |^2 = \Big| [1 + (M + 1)P_s]^2 - P_s^2 | C_M(\Delta\gamma) |^2 - 2P_s[1 + (M + 1)P_s] \text{ Real}\left\{C_M\left(\frac{\Delta\gamma}{2}\right)\right\} + 2P_s^2 \text{ Real}\left\{C_M(\Delta\gamma)C_M^*\left(\frac{\Delta\gamma}{2}\right)\right\} \Big|^2 \quad (5.110)$$

This is solved for $\Delta\gamma = 2\pi \mid f_1 - f_2 \mid \Delta t$ with $C_M(x)$ defined as

$$C_M(x) = \sum_{m=0}^{M} \exp(jx^m) = \frac{\{\exp[jx(M + 1)] - 1\}}{[\exp(jx) - 1]} \quad (5.111)$$

This criterion for resolution R at the minimum $\Delta f = f_2 - f_1$ is independent of centerfrequency f_c. The criterion was derived for complex autocorrelation coefficients, and another analysis of Marple proved that superresolution curves obtained by solving Eq. (5.110) for various P_s and M will represent the mean normalized superresolution over all possible center frequencies of real-valued sinusoids at f_1 and f_2.

Figure 5.3 shows solid-curve AR solutions of Eq. (5.110) for $\Delta\gamma$ for a range of $P_s \equiv$ SNR as a function of M, the order of the ME spectral estimate. Shown for comparison are resolutions R for (1) the Blackman-Tukey, Hanning window, (2) the Fast Fourier Transform Periodogram, which is the same as for the Blackman-Tukey, Rectangular window, and (3) the Blackman-Tukey, Triangular window. Notice the AR or ME resolution of $R < 3.0$, less than half the conventional limit of 2π, is attained for SNR = 0 dB, or signal power equal to noise power in each sinusoid!

The family of AR curves is represented closely for SNR $\geq$ 0 dB by the empirical expression

$$R = 6.47 [(SNR)(M + 1)]^{-0.31} \approx 6.5 (M + 1)^{-1/3} (SNR)^{-1/3} \quad (5.112)$$

In Eq. (5.112) SNR is a ratio rather than a dB value.

To verify these analytic curves, Marple made an ensemble of ME spectral estimates using computer simulation with the harmonic Burg algorithm on real-valued raw data samples [5.2, Sec. 8.4.2]. This algorithm obtains estimates for the filter coefficients $\{\hat{a}_k\}$, $1 \leq k \leq M$, in Eq. (5.109) by recursion. If uses the Levinson algorithm to obtain $\hat{a}_1$ for $p = 1$; $\hat{a}_1$ and $\hat{a}_2$ for $p = 2$; . . . ; and finally $\hat{a}_1$ through $\hat{a}_M$ for $p = M$, where the reflection coefficient $\hat{k}_p = \hat{a}_p$ each step is based on a least-squares criterion.

In the computations, the center frequency was set to $f_M/2$ to produce results representative of the mean resolution $\Delta f = 1/T$ for all center frequencies from 0 to f_M. Each spectrum analysis was based on a 128-sample realization of the process. The pairs of raw-data sequences contained a wide range of initial phases for each sinusoid to average the effect of phase. The dots in Fig. 5.3 indicate the results of the simulations and validate the analytic computations.

These conclusions about superresolution ($R < 2\pi$) of frequency spectra for the time-frequency domains carry over directly to wavenumber spectra of temporally coherent signals for the space-wavenumber domains, by the analogy discussed in Sec. 5.6. The spatial array counterpart of

$$R = 2\pi M\Delta t\Delta f = M\Delta t\Delta\omega \tag{5.113a}$$

is

$$R' = m\Delta x\Delta k_x \tag{5.113b}$$

for an array along the x axis. The classical uncertainty relation is obtained by setting $M\Delta x = L_x$, $\Delta k = 2\pi/L_x$,

$$R'_{\text{class}} = 2\pi = R_{\text{class}} \tag{5.114}$$

The classical resolution limit $\Delta\theta$ in angle of a plane wave arriving with wavenumber $k_x = k \sin\theta$, $\Delta k_x = k \cos\theta\Delta\theta$ is, with $k = 2\pi/\lambda$,

$$\Delta\theta_{\text{class}} = \frac{\Delta k_x}{k\cos\theta} = \frac{\lambda}{L_x\cos\theta} \qquad (\lambda = \text{wavelength}) \tag{5.115}$$

Resolution is better for waves traveling more broadside to the array, $\cos\theta \cong 1$, as expected: the phase shift per element is changing most rapidly with direction θ.

Figure 5.3 of Marple for superresolution of two sinusoidal signals of nearly equal frequency can be directly interpreted for superresolution of two temporally coherent waves of nearly equal wavenumber k_x at a processing frequency ω_o, provided we have the same signal powers in the spatial waves and the same noise powers. One may verify that a wave Eq. (5.54) with unit amplitude and wavenumber k_x^w and frequency ω_o will, according to Eqs. (5.64), (5.66), and (5.67b), have a spectrum $S_{\omega_o}(k_x) = \delta(k_x^w + k_x)$ of the same amplitude as the frequency spectrum of $\exp(j\omega_o t)$, namely, $\delta(\omega - \omega_o)$. The noise components in the two spectra will be the same provided

$$\langle n^*(t)\, n(t + \tau)\rangle = \delta(\tau) \tag{5.116a}$$

for the time-frequency autocorrelation data processing corresponds to

$$\langle\langle n^*(x, t)\, n(x + \rho_x, t + \tau)\rangle\rangle = \delta(\rho_x)\delta(\tau) \tag{5.116b}$$

for the space-wavenumber processing. The $\langle\ \rangle$ means t-averaging; the $\langle\langle\ \rangle\rangle$, both t and x averaging. This space-wavenumber noise is receiver noise, not directional noise of an incoming wave, and uniform in the plane perpendicular to its direction of propagation.

Such ME spatial array processing of sinusoidal-frequency waves with receiver noise present is one of the better ways to estimate directions of incoming waves. This coherent-wave problem is more difficult than the problem of resolving directions of *incoherent* waves with receiver noise present, discussed next.

5.9.2 Two Incoherent Wavenumber Sources

The question of the best superresolution attainable for incoherent sources by various non-ME algorithms as a function of SNR has been addressed. The superresolution of two equistrength incoherent sources in fractions of a beamwidth, $\Delta\theta_{\text{class}}$ of Eq. (5.115), was studied with five adaptive algorithms. A linear array of Fig. 5.3 with exact $\lambda/2$-element spacing at the data processing frequency, was assumed. The five algorithms were: the Weiner matrix inverse, constrained optimized beam scan (maximum likelihood wavenumber spectral estimator), "thermal noise," Howells-Applebaum, and orthonormal lattice filter algorithm described by M. A. Alam. Snapshot raw data averaging was employed. The operation of these algorithms is briefly summarized; details and references are found in [5.16].

The Weiner matrix inverse algorithm relies on an estimate of the array covariance matrix $\mathbf{M}^c$, where the ijth matrix element between receivers i and j of signal voltages $E_i(t)$ and $E_j(t)$ averaged over many snapshots is $\langle M_{ij}^c\rangle = \langle E_i^* E_j\rangle$. Asterisk (*) denotes complex conjugate, which is meaningful if E_i has a complex-amplitude representation. The N eigenvalues and eigenvectors of $\mathbf{M}^c$ are found, with N elements in the array, and the eigenvalue for receiver noise power is denoted B_o. The ith eigenvector is an $N \times 1$ column matrix $\mathbf{e}_i = \text{col}(e_{i1}, \ldots, e_{iN})$, which displays the relative voltages of this "eigenvector beam" as the array elements. An optimized weight vector $\mathbf{W}_o$ is formed, representing as an $N \times 1$ column matrix an *adapted pattern* equal to a reference mainbeam pattern minus a linear combination of eigenvector patterns as determined by $\mathbf{M}^c$. This adapted pattern *nulls the sources,* and the amplitude of its ith eigenvector component is $(\mathbf{e}_i^\dagger \cdot \mathbf{W}_o)$, the dagger (†) denoting transpose conjugate. The inverse of this adapted pattern is the spatial (wavenumber) spectrum.

The constrained optimized beam scan algorithm gives a maximum likelihood wavenumber spectral estimate in the sense that it passes the received power in a narrow directional band while minimizing the power outside this band, subject to a constraint. The optimized weight vector $\mathbf{W}_o$ previously mentioned estimates the adapted pattern and its inverse, the spatial (wavenumber) spectrum, by the residual power $P_o = \mathbf{W}_o^\dagger \mathbf{M}^c \mathbf{W}_o$. When a gain constraint of $\mathbf{S}_{sv}^t \mathbf{W}_o$ is imposed, $\mathbf{S}_{sv}$ being the $N \times 1$ reference mainbeam "steering" vector which displays the array-element excitation for the mainbeam, P_o becomes

$$P_0 = [\mathbf{S}_{sv}^t (\mathbf{M}^c)^{-1} \mathbf{S}_{sv}^*]^{-1} \tag{5.117}$$

$\mathbf{S}_{sv}$ is varied and the spatial spectrum is studied by the eigenvector-beam content of $\mathbf{W}_o = P_o (\mathbf{M}^c)^{-1} \cdot \mathbf{S}_{sv}^*$. When $\mathbf{S}_{sv}$ determines, via $\mathbf{W}_o$, the position of a source(s) in the adapted pattern P_o tends to maximize.

The "thermal noise" algorithm uses a different criterion related to the maximization of P_o for converging on the spatial wavenumber spectrum. When $\mathbf{S}_{sv}$ locates source(s) in the adapted pattern, $|W_o|$ tends to zero. Therefore, the "adapted thermal noise power" $N_o \propto \mathbf{W}_o^\dagger \mathbf{W}_o$ approaches zero and signifies an $\mathbf{S}_{sv}$ vector which is nulling sources in the adapted pattern.

The Howells-Applebaum algorithm updates recursively in real time all the array-element weights in $\mathbf{W}_o(n)$ for a given snapshot n of all the element voltages $\{E_k(n)\}$ from $\mathbf{W}_o(n - 1)$ and the steering vector $\mathbf{S}_{sv}(n)$. The iterations are stopped when the snapshot power ratio $B_o^{-1} \sum_{k=1}^{N} |E_k(n)|^2$ remains acceptably small in a succession of snapshots. As the SNR decreases the spatial wavenumber resolution of the sources deteriorates.

The orthonormal lattice filter algorithm [5.17] is based on a dynamic real-time filter which computes parameters for the spatial AR wavenumber spectrum Eq. (5.82). The numerator is replaced by a smoothed spectral density estimate of the AR filtered error vector, decomposed into a set of orthogonal harmonic vectors. The array $\{a_k\}$ is replaced by estimators $\{\hat{a}_k\}$.

The best superresolution of two equistrength incoherent sources in fractional beamwidth versus SNR obtained by these algorithms is depicted by the dashed line in Fig. 5.4 [5.16]. Each algorithm operated on a series of array snapshot data generated incoherently, so that the voltage $E_k(n)$ at the kth element in the nth snapshot was

$$E_k(n) = w_k(n) + \sum_{i=1}^{2} A_i \exp\{j[k_{xi}x_k + \phi_{ik}(n)]\} \tag{5.118}$$

$\phi_{ik}(n)$ being a random phase in the ith wave.

Resolution of more than two incoherent sources is relatively easy if no more than two sources occupy a beamwidth and the SNR is sufficient according to Fig. 5.4. With three sinusoids within a beamwidth an algorithm may require many snapshots to locate them and their peaks may be ambiguous.

Since the angular resolution improves with SNR, a method of reducing the effect of the noise would increase the effective SNR and reduce the required signal power S for the same resolution. This would reduce the actual SNR for that resolution. Therefore, if one subtracts the receiver noise component of the autocorrelation function $r(\rho_x, \tau)$ in Eq. (5.64) this should improve the ME angular resolution. The broken-line series of x's in Fig. 5.4 shows the improvement to be expected for the resolution of two incoherent sources, using only eight antenna elements spaced $\lambda/2$ apart at the processing frequency, and averaging the raw data over 300 snapshots [the T averaging in Eq. (5.64)] [5.18]. The reduction in SNR required for resolution of two

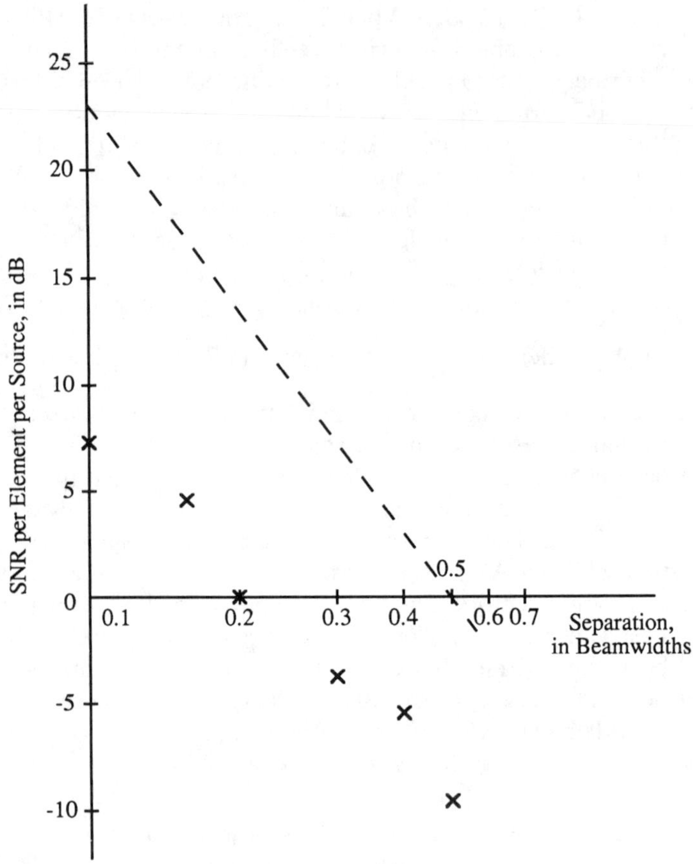

Figure 5.4 A universal approximate resolution limit for two equistrength incoherent sources, with narrowband Gaussian receiver noise [5.16]. Crosses show the expected reduction in required SNR when the noise component is removed from the autocorrelation function [5.18], as described in the text. From [5.15] Fig. 1, © 1982, IEEE.

incoherent sources at various beamwidth separations varied between 7 to 15 dB.

REFERENCES

[5.1] D. Slepian, On bandwidth, Proc. IEEE, *64*, March, 1976, pp. 292–300.

[5.2] S.L. Marple Jr., *Digital Spectral Analysis with Applications,* Englewood Cliffs, NJ: Prentice Hall, 1987.

[5.3] *Programs for Digital Signal Processing,* ed. by the Digital Signal Processing

Committee of the IEEE Acoustics, Speech, and Signal Processing Society, IEEE Press, 1979. Sec. 1.4, "Mixed Radix Fast Fourier Transforms," by R.C. Singleton, and Sec. 1.5, "Optimized Mass Storage FFT Program," by D. Fraser.

[5.4] R.N. McDonough, Maximum entropy spatial processing of array data, *Geophysics, 39*, pp. 843–851, Dec., 1974. Reprinted in *Modern Spectrum Analysis,* D. G. Childers, (ed). IEEE Press, 1978, pp. 72–80.

[5.5] E.T. Jaynes, On the rationale of maximum entropy methods, Proc. IEEE, *70*, Sept., 1982, pp. 939–952.

[5.6] R.N. McDonough, Application of the maximum likelihood method and the maximum entropy method to array processing, *Topics in Applied Physics, 34, Nonlinear Methods of Spectral Analysis,* ed. by S. Haykin, pp. 181–244, Heidelberg: Springer-Verlag, 1983.

[5.7] J.P. Burg, Maximum entropy spectral analysis, In *Modern Spectrum Analysis,* D.G. Childers, (ed). IEEE Press, 1978, pp. 34–41.

[5.8] J. Makhoul, Linear prediction: A tutorial review, Proc. IEEE, *63*, pp. 561–580, Apr., 1975; correction in Proc. IEEE, *64*, p. 285, Feb., 1976. Reprinted in *Modern Spectrum Analysis,* D.G. Childers, (ed.). IEEE Press, 1978, pp. 99–118.

[5.9] S.M. Kay, *Modern Spectral Estimation: Theory and Application,* Englewood Cliffs, NJ: Prentice Hall, 1988, Sec. 15.7.

[5.10] E. Rietsch, The maximum entropy approach to inverse problems, *Jour. Geophys., 42*, 1977, pp. 489–506.

[5.11] N.I. Akhiezer, *Theory of Approximation,* New York, NY: Ungar, 1956.

[5.12] J.A. Edward and M.M. Fitelson, Notes on maximum-entropy processing, IEEE Trans. Information Theory, IT-19, pp. 232–234, Mar., 1973. Reprinted in *Modern Spectrum Analysis,* D.G. Childers, (ed.). IEEE Press, 1978, pp. 94–96.

[5.13] W.D. Smith and D.L. Lager, Evaluation of simple algorithms for spectral parameter analysis of the electroencephalogram, IEEE Trans. Biomed. Eng., *BME-33*, March, 1986, pp. 352–358.

[5.14] Private conversation with W.D. Smith of the Biomedical Engineering Program, California State University, Sacramento, CA, 95819.

[5.15] S.L. Marple, Jr., Frequency resolution of fourier and maximum entropy spectral estimates, *Geophys., 47*, Sept. 1982, pp. 1303–1307.

[5.16] W.F. Gabriel, Spectral analysis and adaptive array superresolution techniques, Proc. IEEE, *68*, June, 1980, pp. 654–666.

[5.17] M.A. Alam, Adaptive spectral estimation, in *Modern Spectrum Analysis.* D.G. Childers (ed.). IEEE Press, 1978, pp. 203–210.

[5.18] A.K. Luthra and B.D. Steinberg, The maximum entropy image of two targets, Proc. IEEE, *70*, Jan., 1982, pp. 98–99.

6

Maximum Entropy Coherent Spectral Analysis (Boltzmann Method)

6.1 INTRODUCTION

In this chapter we develop a lesser-known Boltzmann method of obtaining either an underdetermined frequency spectrum from discrete time-domain data or an underdetermined wavenumber (i.e., propagation direction) spectrum from spatial-domain data. Whereas, the Gibbs's method of Chap. 5 yielded an ME probability distribution for the extended time or spatial array data, and from it, the autocorrelation function and power spectrum, the Boltzmann method yields a frequency or wavenumber complex field amplitude spectrum directly from the available temporal or spatial data. The latter method predicts phases of the complex field amplitudes and is sometimes more appropriate for more grossly underdetermined spectra. That is, the Boltzmann method can exhibit greater superresolution that the Gibbs's method. However, the Boltzmann method does not locate a global maximum of the entropy of the power spectrum.

After summarizing from Chap. 5, the necessary data sampling rules and Fourier transform relations for time-frequency data, which translate by analogy to space-wavenumber data according to Table 5.1, we outline the Boltzmann method. We describe an algorithm which has been implemented to solve various noiseless and noisy synthetic data problems. Noise cannot be entirely removed from short coherent data trains and it gracefully degrades the superresolution with decreasing signal-to-noise ratio. Several examples are given of Boltzmann superresolution of noiseless coherent temporal and spatial array data. In the last section we derive the generalized-inverse relation between data and ME spectrum, state the necessary matrix-rank condition for an inverse, and quote the algorithmic relation between the variance of a spectral amplitude in terms of the noise variances in the data, neglecting the statistical uncertainty component.

Perhaps the primary advantages of the Boltzmann method are that (1) it does not require an underlying probability distribution for an ensemble of data samples, (2) it obtains the ME complex amplitude spectra directly from the data rather than from autocorrelation functions, and (3) the real time-domain data need not be processed at equal intervals of time and the complex-amplitude spatial array data need not be recorded at equal spatial intervals in one, two, or three dimensions.

Boltzmann spectral analysis of incoherent electromagnetic field data is discussed in Chap. 7.

6.2 DATA SAMPLING RULES; FOURIER TRANSFORM RELATIONS

The sampling relations for data in the time domain and Fourier-transform frequency domain, and for the analogous space and Fourier-transform wave-number domains, have been discussed in Chap. 5. We now summarize the pertinent relations for this chapter.

Consider the following spectral resolution problem for data in the time and frequency domains: a temporal waveform $y(t)$ is recorded over the range $(-T, +T)$[1] and defined at intervals of Δt, where Δt is related to the bandwidth over $(-W, +W)$ by the Nyquist sampling relation $\Delta t = 1/(2W)$. Either the signal bandwidth is actually so limited by the physical process involved or the measurement scheme or else the *effective* bandwidth is set by Δt itself. The problem is to resolve the complex frequency spectrum $g(f)$ at prescribed intervals over $(-W, +W)$. Since $y(t)$ is real, $g(-f) = g^*(f)$ [the complex conjugate of $g(f)$] and we have only to resolve $g(f)$ over $0 \leqq f \leqq W$. Ordinary resolution by the sampling relation dictates that the frequency resolution should be $\Delta f = 1/(2T)$. If so, the number of positive time points $N = T/\Delta t$ equals the number of positive frequency points $M = W/\Delta f$ and ordinary Fast Fourier Transform (FFT) techniques will yield the complex amplitude spectrum at intervals of Δf.

Suppose, now, we wish to estimate the spectrum at finer intervals $\Delta f' < \Delta f$, as if we knew $y(t)$ over a range wider than $(-T, +T)$. The data $y(t)$ are related to $g(f)$ formally by the Fourier transform,

$$y(t) = \frac{1}{2\pi}\int g(\omega)e^{j\omega t}d\omega, \qquad \omega = 2\pi f \tag{6.1}$$

[1] This differs from the notation in Chap. 5 where the measurement interval was (0,T).

and if we specify $g(\omega)$ at intervals $f_m = m\Delta f'$ for $-M \leqq m \leqq M$, where now, $M > N = T/\Delta t$, then the bandlimited $y(t)$ is determined at the intervals t_n for $n\Delta t$ for $-N \leqq n \leqq N$ by the discrete version of Eq. (6.1),

$$y_n = y(t_n) = \frac{\Delta\omega'}{2\pi} \sum_{m=-M}^{M} g_m \exp(m\Delta\omega' n\Delta t), \qquad \begin{cases} \Delta\omega' = 2\pi\Delta f' \\ -N \leq n \leq N. \end{cases} \tag{6.2}$$

We simplify the exponent by substituting $\Delta f' = W/M$, $\Delta t = 1/(2W)$, so it reads

$$m\Delta\omega' n\Delta t = \frac{\pi mn}{M} \tag{6.3}$$

Defining the weighted discrete spectrum G_m by $g_m\Delta\omega/2\pi$, we write Eq. (6.2) as

$$y_n = \sum_{m=-M}^{M} G_m e^{j\pi mn/M}, \qquad -N \leq n \leq N < M \tag{6.4}$$

This is now a set of $(2N + 1)$ constraint relations for the $(2M + 1)$ complex values of weighted spectrum G_m. Since $G_{-m} = G_m^*$ there are $(2M + 1)$ real spectral unknowns in all the G_m amplitudes.

We do not utilize the Fourier transform dual to Eq. (6.4) for each G_m as a weighted sum of all the y_n because the M values of G_m are improperly determined by the $N < M$ values of the y_n. More y_n data outside the range $(-T,T)$, at intervals of $\Delta t = 1/(2W)$, would be required to determine all the G_m by the transform relation dual to Eq. (6.4).

If we were performing partially coherent or incoherent spectral analysis we could replace Eq. (6.4) by

$$r_n = r(n\Delta t) = \sum_{m=-M}^{M} S_m e^{j\pi mn/M} \tag{6.5}$$

r_n being the real autocorrelation function and S_m the weighted power spectral density at frequency $f_m = m\Delta f'$. This equation was solved by Gibbs's method in Sec. 5.5 [6.1].

6.3 THE OBJECTIVE FUNCTION F FOR THE MAXIMUM ENTROPY FREQUENCY SPECTRUM

Our method of deriving an ME complex amplitude spectrum G_m, with $G_{-m} = G_m^*$, $-M \leqq m \leqq M$, is to write an entropy power spectrum analogous to Eq. (2.26), in which $|G_m|^2$ corresponds to the number of building blocks in the mth frequency cell and $\sum |G_m|^2$, to the total number of building blocks. Since $| G_m |^2 = G_{mR}^2 + G_{mI}^2$, subscripts R and I denoting real and imaginary parts of the coherent amplitude, we have for entropy H,

$$H = \log\left(\frac{[\sum_m (G_{mR}^2 + G_{mI}^2)]!}{\prod_m (G_{mR}^2 + G_{mI}^2)!}\right) = -\sum_m | G_m |^2 \log\left(\frac{| G_m |^2}{\sum_k | G_k |^2}\right) \tag{6.6}$$

This last expression is Shannon's form multiplied by $A_T = \sum | G_k |^2$, which is allowed to vary with the $| G_k |^2$ as the ME solution point in G_m-space is approached.

Because $G_{-m} = G_m^*$ in the frequency spectrum the sum over m runs from 0 to M, with the $m \geqq 1$ terms doubled to account for the negative frequencies. Therefore, Eq. (6.6) for the frequency spectrum reads

$$H = -G_0^2 \log\left(\frac{G_0^2}{A_T}\right) - 2\sum_{m=1}^{M} | G_m |^2 \log\left(\frac{| G_m |^2}{A_T}\right) \tag{6.7a}$$

with

$$A_T = G_0^2 + 2\sum_{m=1}^{M} | G_m |^2 \tag{6.7b}$$

Our objective function F is composed of this H plus the sum of the (2N + 1) constraints Eq. (6.4), each multiplied by an (unknown) Lagrange multiplier β_n. The y_n are all real, and $G_{-m} = G_m^*$, so we can combine the $\pm m$ frequency terms and write F as

$$F = H\,[\text{of Eq. (6.7)}] + \sum_{n=-N}^{N} \beta_n \left\{ G_0 + 2\sum_{m=1}^{M} G_{mR} \cos\left(\frac{\pi mn}{M}\right) - G_{mI} \sin\left(\frac{\pi mn}{M}\right) - y_n \right\} \tag{6.8}$$

By extremizing F for small changes in the G_0, G_{mR}, and independently, the G_{mI}, we obtain (2M + 1) equations for them. The constraint relations for the various n within the braces of Eq. (6.8) furnish another (2N + 1) equation. Self-consistent solution of all these yields the (2N + 1) real β_n and the complex G_m-spectrum which extremizes H of Eq. (6.7) subject to the constraints. If we expand $H(\{(G_{mR} + \delta G_{mR})^2, (G_{mI} + \delta G_{mI})^2\})$ about $\{G_{mR}, G_{mI}\}^{ME}$ through quadratic terms we find that H is not necessarily maximum relative to a change δG_0 in amplitude G_0 and δ-changes in the phases of the G_m for $m > 0$. This is so whether A_T of Eq. (6.7) is constrained or allowed to vary with the $| G_m |^2$. This matter is discussed further for the analogous and more general ME wavenumber spectrum following Eq. (6.16).

We have not constrained the phases $\tan^{-1}(G_{mI}/G_{mR})$ at all, yet phase is predicted remarkably accurately in all the synthetic data examples we have impartially selected for study. Some of these are described in Sec. 6.6. A_T, representing the total amount of spectral power, is usually not known so it was allowed to "float." In our algorithm it was obtained relative to a parameter $A'_T > A_T$, which was changed each iteration of a procedure which converged to the self-consistent $\{\beta'_n\}$ and $\{G_{mR}, G_{mI}\}$. The entropy is maximized relative to A'_T and the $\{\beta_{nR}, \beta_{nI}\}$ each iteration, so as to satisfy the constraints and A'_T is made to approach A_T as the self-consistent solution is approached.

Unfortunately, there appears to be neither a rigorous nor an empirical formula for the Boltzmann spectrum analogous to Eq. (5.19) for the Burg spectrum, in terms of filter coefficients found by solving the autocorrelation matrix equation Eq. (5.18). After we obtain the objective function for the ME wavenumber spectrum in Sec. 6.4, we will outline in Sec. 6.5 an algorithm for converging on either spectrum. This algorithm could also serve to track changes in the ME G_m values due to changes in the data, as was done in Sec. 3.8, Eq. (3.58), for geotomographic problems. The matrix formulas for these changes are found in Sec. 6.7.

6.4 THE OBJECTIVE FUNCTION F FOR THE MAXIMUM ENTROPY WAVENUMBER SPECTRUM

We describe an array of electromagnetic dipoles (short compared to the wavelength $\lambda = c/f$, c being light velocity, at the processing frequency $\omega = 2\pi f$) which act as receivers aligned as shown in Fig. 6.1. The nth receiver located at spherical coordinates (r_n, θ_n, ϕ_n) is oriented in vector direction $\bar{a}_n$ and records a net coherent signal of complex amplitude Y_n from all the distant radiators. The mth radiator at (R_0, θ_m, ϕ_m) is oriented in direction $\bar{a}_m$ and has complex amplitude G_m. We want to estimate the various (G_m, θ_m, ϕ_m) from the array $\{Y_n\}$.

That part of Y_n received from G_m is

$$\bar{a}_n \cdot \bar{a}_m G_m \frac{e^{-j\omega R_{mn}/c}}{4\pi R_{mn}}, \quad c = \text{light velocity} \tag{6.9}$$

For $R_0 \gg$ each r_n,

$$R_{mn} \approx \begin{cases} R_0 - r_n[\cos\theta_m \cos\theta_n + \sin\theta_m \sin\theta_n \cos(\phi_m - \phi_n)] & \text{in the exponent} \\ R_0 & \text{in the denominator} \end{cases} \tag{6.10}$$

We regard all sources as effectively at a distance R_0 and absorb the common factor $\exp(-j\omega R_0/c)/4\pi R_0$ into each Y_n. Upon adding all the G_m contribu-

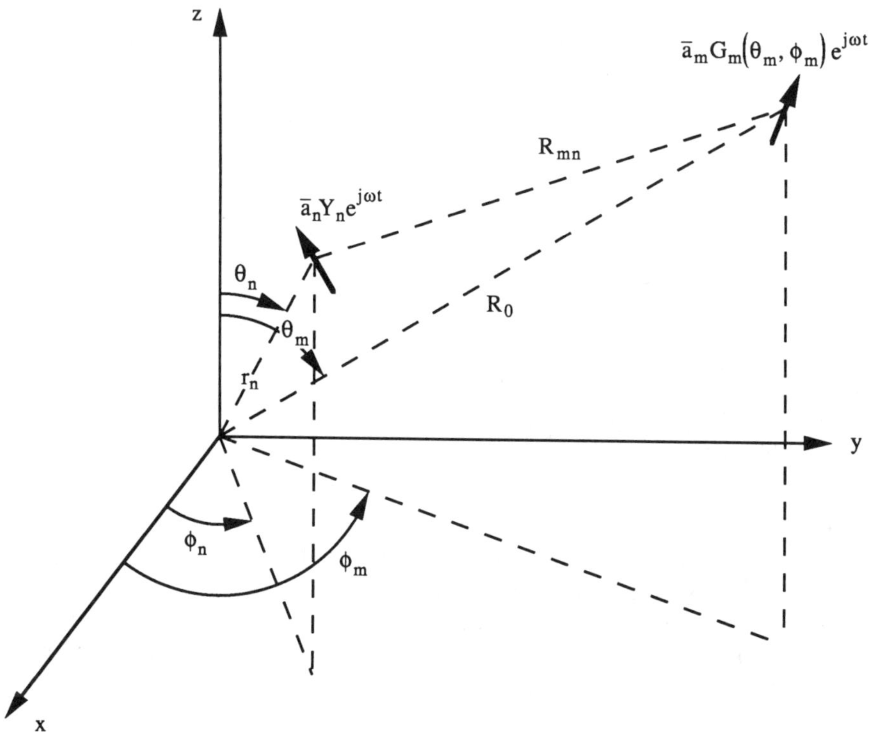

Figure 6.1 An array of dipole receivers of electromagnetic radiation at frequency ω ($\exp j\omega t$ time variation) from a distribution of point far-field sources. Receiver n is at spherical coordinates (r_n, θ_n, ϕ_n), oriented in direction $\bar{a}_n$ and records a signal of complex amplitude Y_n. Source m has a dipole moment $\bar{a}_m G_m$ in the far-field at (R_o, θ_m, ϕ_m). $R_{mn} \gg \lambda = 2\pi c/\omega$.

tions to Y_n we have

$$Y_n \approx \bar{a}_n \cdot \sum_{m=1}^{M} \bar{a}_m\, G_m \times \exp\left\{+j\frac{\omega}{c}r_n\left[\cos\theta_m\cos\theta_n + \sin\theta_m\sin\theta_n\cos(\phi_m - \phi_n)\right]\right\} \tag{6.11}$$

This is analogous to Eq. (6.4) for the time series, except that Y_n is complex and affected by the vector coupling between the nth receiver and all the possible radiators. The complex-amplitude G_m-sources are all treated as independent of each other.

The objective function F is now h of Eq. (6.6) plus all the N constraints

Eq. (6.11), each times a Lagrange multiplier. It is expedient to split Eq. (6.11) into real and imaginary parts, each multiplied by a real β_{nR} and β_{nI}, respectively. Then F is

$$F = h[\text{of Eq. (6.6)}] + \sum_{n=1}^{N} \beta_{nR}\left[-Y_{nR} + \bar{a}_n \cdot \sum_{m=1}^{M} \bar{a}_m(G_{mR}C_{mn} - G_{mI}S_{mn})\right]$$
$$+ \sum_{n=1}^{N} \beta_{nI}\left[-Y_{nI} + \bar{a}_n \cdot \sum_{m=1}^{M} \bar{a}_m(G_{mR}S_{mn} + G_{mI}C_{mn})\right] \tag{6.12}$$

with

$$\begin{matrix} C_{mn} \\ S_{mn} \end{matrix} = \begin{matrix} \cos \\ \sin \end{matrix}\left\{\frac{\omega}{c} r_n \left[\cos\theta_m \cos\theta_n + \sin\theta_m \sin\theta_n \cos(\phi_m - \phi_n)\right]\right\} \tag{6.13}$$

This is the general expression for arbitrarily oriented receivers and radiators in Fig. 6.1.

Subsequently, we will consider synthetic data examples of a simpler configuration, with all the receivers in the y − z plane and oriented parallel to the x axis: all $\bar{a}_n = \bar{a}_x$, $\phi_n = \pi/2$. These will locate any mth source with $\bar{a}_m \cdot \bar{a}_n \neq 0$ at any (θ_m, ϕ_m). To simplify further we shall assume all sources are in the $y - z$ plane, all $\phi_m = \pi/2$, at various θ_m. Then, Y_n simplifies to

$$Y_n = \sum_{m=1}^{M} G_m \exp\left[j\frac{\omega}{c} r_n \cos(\theta_m - \theta_n)\right], \qquad 1 \le n \le N \tag{6.14}$$

and F becomes

$$F = h[\text{of Eq. (6.6)}] + \sum_{n=1}^{N} \beta_{nR}\left[\sum_{m=1}^{M} (G_{mR}C_{mn} - G_{mI}S_{mn}) - Y_{nR}\right]$$
$$+ \sum_{n}^{N} \beta_{nI}\left[\sum_{m}^{M} (G_{mR}S_{mn} + G_{mI}C_{mn}) - Y_{nI}\right] \tag{6.15}$$

where, now,

$$\begin{matrix} C_{mn} \\ S_{mn} \end{matrix} = \begin{matrix} \cos \\ \sin \end{matrix}\left[\frac{\omega}{c} r_n \cos(\theta_m - \theta_n)\right] \tag{6.16}$$

Although H of Eq. (6.6) is an extremum at the ME point $\{G_{mR}, G_{mI}\}^{ME}$ it is not a global maximum. To understand this, we first observe the convex property of H relative to any $G_k = \sqrt{G_{kR}^2 + G_{kI}^2}$. We distinguish two cases: (i) $A_T = \sum G_k^2$ constrained, for which $\partial H/\partial G_k = 2G_k(|L_k| - 1)$ implies H has a local maximum at $A_T/G_k^2 = e$, or $G_k^2/A_T = 0.368$. The "entropy bowl," an initially convex plot of $H(G_k)$, changes slope at this value of G_k^2/A_T. For case (ii), A_T unconstrained, $\partial H/\partial G_k = 2G_k \mid L_k \mid \geq 0$, and H has no local maximum relative to $\mid G_k^2 \mid /A_T < 1$. The entropy bowl is strictly convex upward.

We can study the properties of H at an ME point by expanding H $(\{(G_{mR} + \delta G_{mR})^2, (G_{mI} + \delta G_{mI})^2\})$ through all terms quadratic in the δG_{mR} and δG_{mI}. This is done by evaluating $\partial H/\partial(G_{mR}^2)$, $\partial H/\partial(G_{mI}^2)$, $\partial^2 H/\partial(G_{mR}^2)^2$, $\partial^2 H/\partial(G_{mI}^2)^2$, and $\partial^2 H/\partial G_{mR}^2 \partial G_{mI}^2$. The linear δG_{mR} and δG_{mI} terms in ΔH sum over m to zero because the constraints must be satisfied. The quadratic terms give

(i) $A_T = \sum \mid G_k \mid^2$ constrained,

$$\Delta H, \text{quadratic}, = \sum(\Delta H)_k = \sum_k \left\{ -(L_k + 1)[(\delta G_{kR})^2 + (\delta G_{kI})^2] - 2\left[\frac{G_{kR}\delta G_{kR}}{\mid G_k \mid} + \frac{G_{kI}\delta G_{kI}}{\mid G_k \mid}\right]^2 \right\}$$

$$= \sum_k \left\{ (\mid L_k \mid - 1) - 2\cos^2 \psi_k \right\} \mid \delta G_k \mid^2$$

where $L_k = \log(\mid G_k \mid^2/A_T) < 0$, $\mid \delta G_k \mid^2 = (\delta G_{kR})^2 + (\delta G_{kI})^2$.

(ii) A_T unconstrained,

$$\Delta H, \text{quadratic}, = \sum(\Delta H)_k = \sum_k \left\{ -L_k[(\delta G_{kR})^2 + (\delta G_{kI})^2] - 2\left(1 - \frac{\mid G_k \mid^2}{A_T}\right)\left[\frac{G_{kR}\delta G_{kR}}{\mid G_k \mid} + \frac{G_{kI}\delta G_{kI}}{\mid G_k \mid}\right]^2 \right\}$$

$$= \sum_k \left\{ \mid L_k \mid - 2\left(1 - \frac{\mid G_k \mid^2}{A_T}\right) \cos^2\psi_k \right\} \mid \delta G_k \mid^2$$

The final summands previously mentioned are obtained by writing

$$\left[\frac{G_{kR}\delta G_{kR}}{\mid G_k \mid} + \frac{G_{kI}\delta G_{kI}}{\mid G_k \mid}\right] = \bar{1}_k \cdot \overline{\delta G_k} = \mid \overline{\delta G_k} \mid \cos \psi_k$$

and ψ_k is defined in the vector sketch on page 122.

Regions of $(\Delta H)_k < 0$ and > 0 are shown, symmetric about the $\psi_k = \pm 90°$ line.

If $\mid G_k \mid^2$ is relatively small and $\mid L_k \mid$ relatively large the { }-expression in either $(\Delta H)_k$ of interest will be >0 for all ψ_k. If $A_T/\mid G_k \mid^2$ has an inter-

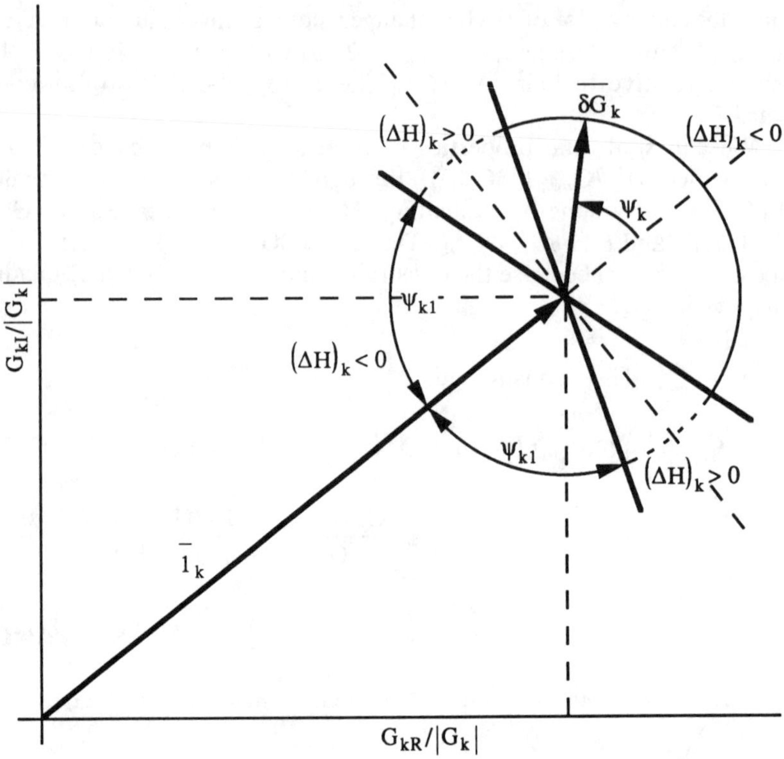

mediate value there are two ψ_{k1} lines of $(\Delta H)_k = 0$, as shown. For case (i), as $A_T/|G_k|^2 \rightarrow e$ from above, $|L_k| \rightarrow 1+$ and $\psi_{k1} \rightarrow 90°$. If $A_T/|G_k|^2 \leq e$, $(\Delta H)_k \leq 0$ for all ψ_k. For case (ii) there is always a ψ_{k1} "saddle" for $A_T/|G_k|^2 < 5.0$.

Our method of extremizing F of Eq. (6.8) subject to constraints is "maximum" entropy in the sense that H is maximum for a change of a relatively large amplitude $|G_k|$, as measured by $|G_k|^2/A_T$, in either constant-phase direction, although it usually is a minimum in either $\psi_k = \pm 90°$ direction of changing phase. For case (i), if at lease one $|L_k|$ is >1 we can specify an infinity of perturbations $\{\delta G_k\}$ in the $\psi_k = \pm 90°$ directions so as to make $\Delta H > 0$.

However, numerical examples show that all the 2N phase constraints of Eq. (6.14) often force the phases of the *large* G_k^{ME} to be very accurate, and the ME-amplitude spectrum often resolves very well a true point-source spectrum.

It is conceivable that more than one ME solution for $G = \{G_{mR}, G_{mI}\}$ as given by Eq. (6.43) could exist in a problem. If so, we would select that solution of greatest H.

6.5 AN ALGORITHM FOR THE ME G_m-AMPLITUDE SPECTRUM

6.5.1 Frequency Spectrum

This iterative algorithm makes the objective function F of Eq. (6.8) for the frequency spectrum extremal with respect to variations δG_{mR}, δG_{mI} of the spectral amplitudes.

From F of Eq. (6.8), we obtain the following equations for the real G_0, G_{mR}, and G_{mI} for $1 \leqq m \leqq M$, leaving $A_T = G_0^2 + 2 \sum | G_m |^2$ unconstrained,

$$G_0 = \frac{1}{2L_0} \sum_{n=-N}^{N} \beta_n \tag{6.17}$$

$$G_{mR} = \frac{1}{2L_m} \sum_n \beta_n \cos\left(\frac{\pi mn}{M}\right) \tag{6.18}$$

$$G_{mI} = -\frac{1}{2L_m} \sum_n \beta_n \sin\left(\frac{\pi mn}{M}\right) \tag{6.19}$$

in which

$$L_0 = \log\left(\frac{G_0^2}{A_T}\right)$$

$$L_m = \log\left[\frac{(G_{mR}^2 + G_{mI}^2)}{A_T}\right] \qquad \text{for } m \geq 1 \tag{6.20}$$

$$A_T = G_0^2 + 2 \sum_{m=1}^{M} (G_{mR}^2 + G_{mI}^2) \tag{6.21}$$

There is an implied constraint obtained from all the (2N + 1) constraints within the parentheses of Eqs. (6.8) and (6.17) to (6.20) which proves useful in scaling the G_m,

$$-h = G_0^2 L_0 + 2 \sum_{m=1}^{M} (G_{mR}^2 + G_{mI}^2) L_m = \frac{1}{2} \sum_{n=-N}^{N} \beta_n y_n \tag{6.22}$$

Matrix equations for updating the β_n are obtained by multiplying Eq. (6.18) by $2 \cos(\pi mn'/M)$ and summing over m, multiplying Eq. (6.19) by $-2 \sin(\pi mn'/M)$ and summing over m, and then adding all these equations to Eq. (6.17). According to the y_n-constraint we obtain

$$\sum_{n=-N}^{N} F_{n'n}\beta_n = y_{n'}, \qquad -N \leq n' \leq N \tag{6.23}$$

where

$$F_{n'n} = \frac{1}{2L_0} + \sum_{k=1}^{M} \frac{\left[\cos\left(\frac{\pi k n'}{M}\right)\cos\left(\frac{\pi k n}{M}\right) + \sin\left(\frac{\pi k n'}{M}\right)\sin\left(\frac{\pi k n}{M}\right)\right]}{L_k} \tag{6.24}$$

In this algorithm for the frequency spectrum, we make initial estimates of the relative g_m of Eq. (6.2) from the discrete inverse transform dual to Eq. (6.2) but for the *Nyquist interval* $\Delta f = 1/(2T) = W/N$

$$g_m = g(m\Delta f) = \Delta t \sum_{n=-N}^{N} y_n \exp(-jm\Delta\omega n\Delta t)$$

$$= \Delta t \sum_n y_n \exp\left(\frac{-j\pi mn}{N}\right) \qquad \begin{array}{l}\Delta\omega = 2\pi\Delta f \\ -N \le m \le N\end{array} \tag{6.25}$$

and then interpolate to the intermediate frequencies $m\Delta f' = mW/M$ to obtain initial guesses of the $g(m\Delta f')$. These yield the initial $G_m = g(m\Delta f')\Delta\omega'/2\pi$ and the L_0, L_m's of Eq. (6.20), followed by the $F_{n'n}$ of Eq. (6.24) and the initial β_n by inversion of Eq. (6.23).

6.5.2 Wavenumber Spectrum

This version of the iterative algorithm makes the objective function F of Eq. (6.15) extremal relative to variations δG_{mR}, δG_{mI} of the wavenumber spectrum. The result, analogous to Eqs. (6.17) to (6.19) with $A_T = \sum |G_k|^2$ unconstrained, is

$$G_{mR} = \frac{1}{2L_m} \sum_{n=1}^{N} (\beta_{nR} C_{mn} + \beta_{nI} S_{mn}) \tag{6.26}$$

$$G_{mI} = \frac{1}{2L_m} \sum_n (-\beta_{nR} S_{mn} + \beta_{nI} C_{mn}), \qquad 1 \le m \le M \tag{6.27}$$

in which

$$L_m = \log\left[\frac{(G_{mR}^2 + G_{mI}^2)}{A_T}\right] \tag{6.28a}$$

$$A_T = \sum_{m=1}^{M} (G_{mR}^2 + G_{mI}^2) \tag{6.28b}$$

The implied constraint analogous to Eq. (6.22) is obtained from Eqs.

(6.14) and (6.26) to (6.28) as

$$-H = \sum_{m=1}^{M} (G_{mR}^2 + G_{mI}^2)\, L_m = \frac{1}{2} \sum_{n=1}^{N} (\beta_{nR} Y_{nR} + \beta_{nI} Y_{nI}) \qquad (6.29)$$

This proves useful in scaling the G_m.

Matrix equations for the β_{nR}, β_{nI} are obtained by multiplying Eq. (6.26) by $C_{mn'}$, Eq. (6.27) by $S_{mn'}$, subtracting or adding one to the other, and then summing over m. These operations yield for the real and imaginary parts of Eq. (6.14)

$$\sum_{n=1}^{N} (F_{n'n}\beta_{nR} + G_{n'n}\beta_{nI}) = Y_{n'R} \qquad n' \leq N \qquad (6.30)$$

$$\sum_{n} (-G_{n'n}\beta_{nR} + F_{n'n}\beta_{nI}) = Y_{n'I} \qquad (6.31)$$

in which the $F_{n'n}$ and $G_{n'n}$ coefficients are defined as

$$F_{n'n} = \frac{1}{2} \sum_{k=1}^{M} \frac{(C_{kn}C_{kn'} + S_{kn}S_{kn'})}{L_k} \qquad (6.32)$$

$$G_{n'n} = \frac{1}{2} \sum_{k} \frac{(S_{kn}C_{kn'} - C_{kn}S_{kn'})}{L_k} \qquad (6.33)$$

C_{kn} and S_{kn} are given by Eq. (6.16).

In the wavenumber-spectrum algorithm we can take all the initial L_m of Eq. (6.28) to be equal, and the initial β_{nR} and β_{nI} follow from inversion of Eqs. (6.30) and (6.31) once the $F_{n'n}$ and $G_{n'n}$ are computed.

6.5.3 Algorithm Details

The algorithm for either version of spectral analysis is iterated through a loop in which the initial $\{\beta_{nR}, \beta_{nI}\}$, computed as just described, determine the amplitude spectrum $\{G_m\}$ by Eqs. (6.17) to (6.21) or (6.26) to (6.28), which are then scaled to satisfy Eqs. (6.22) or (6.29), after which the predicted data $\{\hat{y}_n\}$ of Eq. (6.4) or $\{\hat{Y}_n\}$ of Eq. (6.14) are computed. If these are too far from the actual data $\{y_n\}$ or $\{Y_n\}$, the $\{\beta_n\}$ are recomputed by matrix inversion of Eqs. (6.23) or (6.30) to (6.31) and the loop is repeated.

Typical synthetic data problems of the type described in Sec. 6.6 require ten or so iterations on a CDC 7600 computer to achieve several percent accuracy in the ME G_m spectrum.

A facilitating "entropy shift" is discussed based on an A_T parameter used in place of A_T in Eqs. (6.21) or (6.28). We discuss this for the frequency spectrum data processing; the remarks apply by analogy to the wavenumber data processing. The need for an $A_T' \geq A_T$ arises from considerations of the

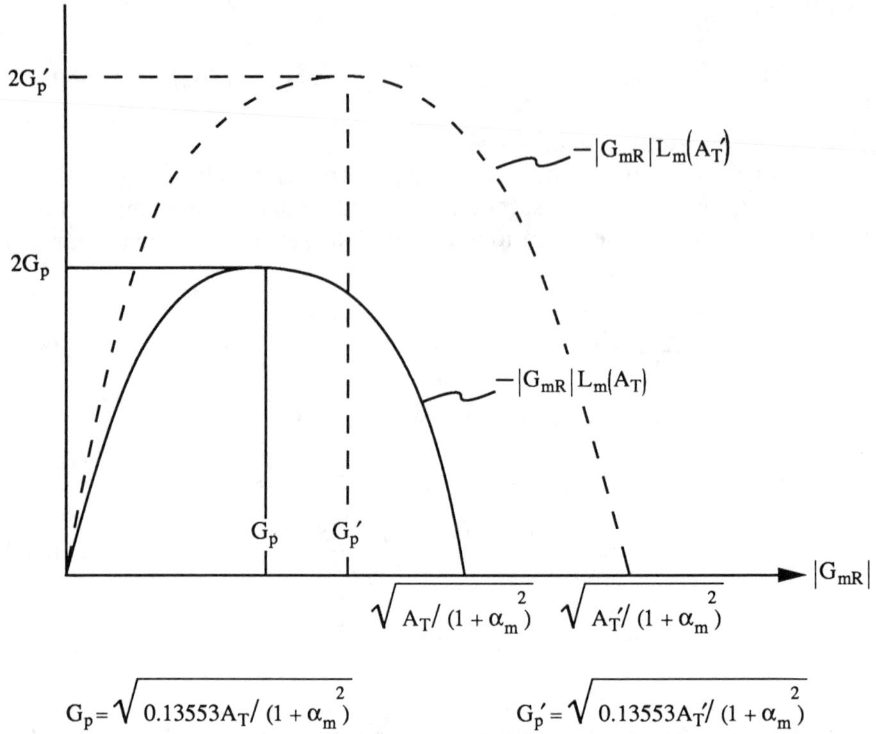

Figure 6.2 Behavior of the function $|G_{mR}|L_m$ of Eq. (6.18) or (6.34) for parameters A_T and $A'_T > A_T$ in Eq. (6.21) or (6.28). $\alpha_m = G_{mI}/G_{mR}$.

behavior of Eqs. (6.17) and (6.18) for a current value of A_T in Eq. (6.21); see Fig. 6.2. If we divide Eq. (6.19) by Eq. (6.18), we obtain a value for the ratio $\alpha_m = G_{mI}/G_{mR}$. Substituting this into Eq. (6.18) it becomes

$$G_{mR}L_m = G_{mR}\log\left(\frac{G_{mR}^2(1+\alpha_m^2)}{A_T}\right)$$

$$= \frac{1}{2}\sum_n(\beta_{nR}C_{mn} + \beta_{nI}S_{mn}) \quad \text{for each } m \tag{6.34}$$

The magnitude of the left side of this is plotted qualitatively versus $|G_{mR}|$ as the solid curve in Fig. 6.2. The rationale for introducing $A'_T > A_T$ is that it enables us to satisfy Eq. (6.34) with A'_T replacing A_T for a wider range of G_{mR} values over the positive-slope portion of the larger dashed curve rather than over the same range of the smaller, double-valued solid curve.

As the iterations proceed, we make $A'_T \to A_T$ from above A_T and $h(A'_T) \to h(A_T)$ in Eq. (6.6), but may discover that one or more of the G_{mR} values do pass to the negative-slope side of the current dashed A'_T curve at one of the iterations. This can be accounted for. The final ME solution has a set of $\{\beta_{nR}, \beta_{nI}\}$ which yields the $\{G_{mR}, G_{mI}\}$ by Eqs. (6.17), (6.19)—with $A'_T = A_T$ in Eq. (6.20)—and also satisfies the constraints Eq. (6.23). According to previous remarks this is our conditionally maximum entropy solution.

6.6 SYNTHETIC DATA EXAMPLES

First we derive the Boltzmann ME spectrum of this noiseless, real-time function

$$y(t) = \cos(2\pi f_o t) \qquad \text{at } f_o = \frac{W}{2} \tag{6.35}$$

where the effective bandwidth $W = 1/(2\Delta t)$ is 2 because we sample at intervals of $\Delta t = 1/4$. Instead of choosing the number of positive time points N and the duration $(-T, +T)$ of the measured signal with $T = N\Delta t$ and then increasing the number of resolved positive frequencies $M > N$ for successively greater superresolution we choose to do just the reverse. We fix the number of positive frequencies at $M = 20$ so $\Delta f' = 2/20 = 0.1$. Then, we reduce $N < M$ for successively greater superresolution. This makes the Nyquist sampling interval $\Delta f = W/N > \Delta f'$.

The relative real part G_{mR} of the positive frequency spectrum is plotted in Fig. 6.3 over $0 \leqslant m \leqslant 20$ for three cases of successively greater superresolution

$$\begin{aligned} N &= 19, \ \Delta f = \frac{W}{N}\ (\text{Nyquist}) = \frac{2}{19} = 0.105, \ T = 4.75 \\ N &= 10, \qquad\qquad\qquad\qquad = \frac{2}{10} = 0.200, \quad\ \ = 2.50 \\ N &= 5, \qquad\qquad\qquad\qquad\ \ = \frac{2}{5} = 0.400, \quad\ \ = 1.25 \end{aligned} \tag{6.36}$$

G_{mR} and G_{mI} versus index m are solutions of Eqs. (6.17) to (6.24); the algorithm of the preceding section yields negligible imaginary part, G_{mI}, of the spectrum. Its magnitude is $\approx 10^{-15}$. The algorithm has not computed G_{mI} to be exactly zero because Eq. (6.35) is an even function of time but the ME result says that a negligible G_{mI} is overwhelmingly probable.

The first case of $N = 19$, slight underdetermination of the spectrum, could be treated by the Gibbs-Burg method of Sec. 5.3 by extending $T = 19\Delta t$ by one time point at each end of the signal to $T' = 20\Delta t$, making $\Delta f' = 1/2T' = 0.1$. This is less than the Nyquist interval $\Delta f = 0.105$ established by $N = 19$. The predicted Boltzmann spectrum for $N = 19$ and no noise

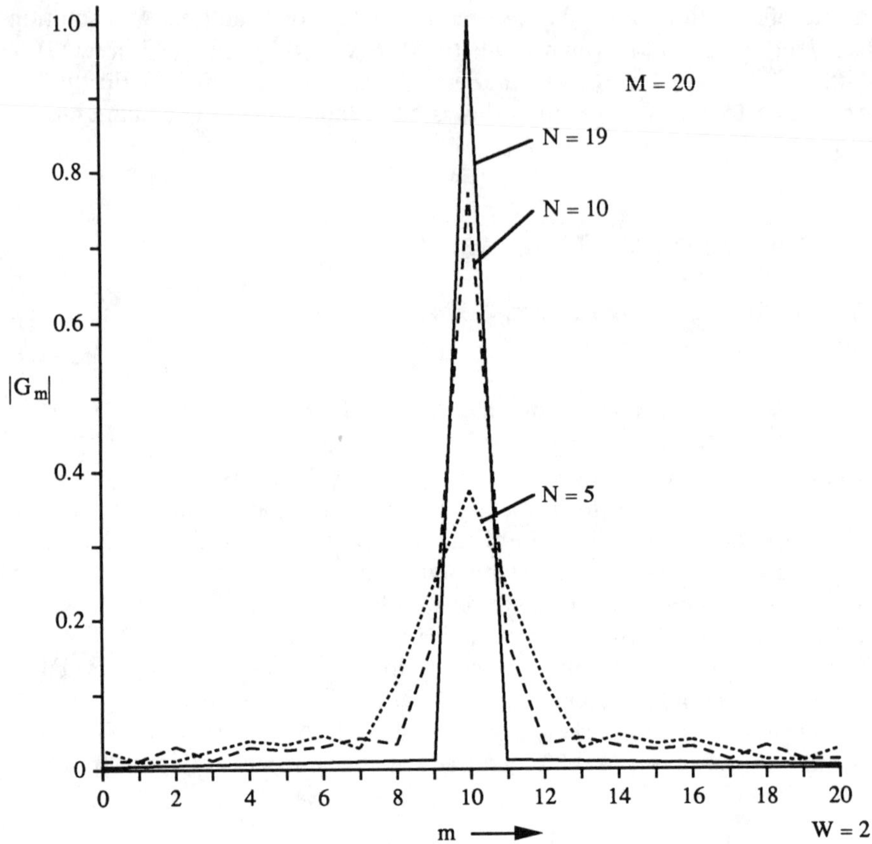

Figure 6.3 The real part, $|G_{mR}|$, of the relative positive-frequency ME spectrum of Eq. (6.35) according to Eqs. (6.17) to (6.19) versus frequency $f = m\Delta f'$. W = 2 and $\Delta t = 1/4$. M = 20 frequencies are resolved with discrimination $\Delta f' = 2/20 = 0.1$. N < M time samples with Nyquist resolution $\Delta f = W/N > \Delta f'$ superresolve the spectrum according to Eq. (6.36). No noise is present.

added strongly resembles in shape the Gibbs-Burg spectrum with 10 percent white noise added, plotted by the dashed line on a dB scale in Fig. 6.4 [6.2].

As N decreases in Fig. 6.3, the spectral resolution degrades because the "object" Eq. (6.35) is measured over shorter durations $T = N\Delta t = N/4$. But even for the extreme case of N = 5 and $\Delta f = 0.4$, *four times* the resolution interval $\Delta f' = 0.1$ for N = M = 20, the ME spectrum still shows a prominent spectral peak at precisely the expected frequency $f_o = W/2$.

On Fig. 6.4 we show three curves of predicted spectral powers, each plotted relative to its own average power (0 dB): the Gibbs-Burg prediction by the method of Sec. 5.3 with 10 percent white noise added to Eq. (6.35) (dashed line), the N = 19 Boltzmann curve of Fig. 6.3 with no noise (broken line), and the N = 10 Boltzmann curve of Fig. 6.3 with no noise (solid line). Ten percent white noise for a unit power sinusoid [6.2] means an average signal power of 0.5/W over the positive-frequency band and an average noise power one-tenth of this. The curves show that the resolution of the Gibbs-

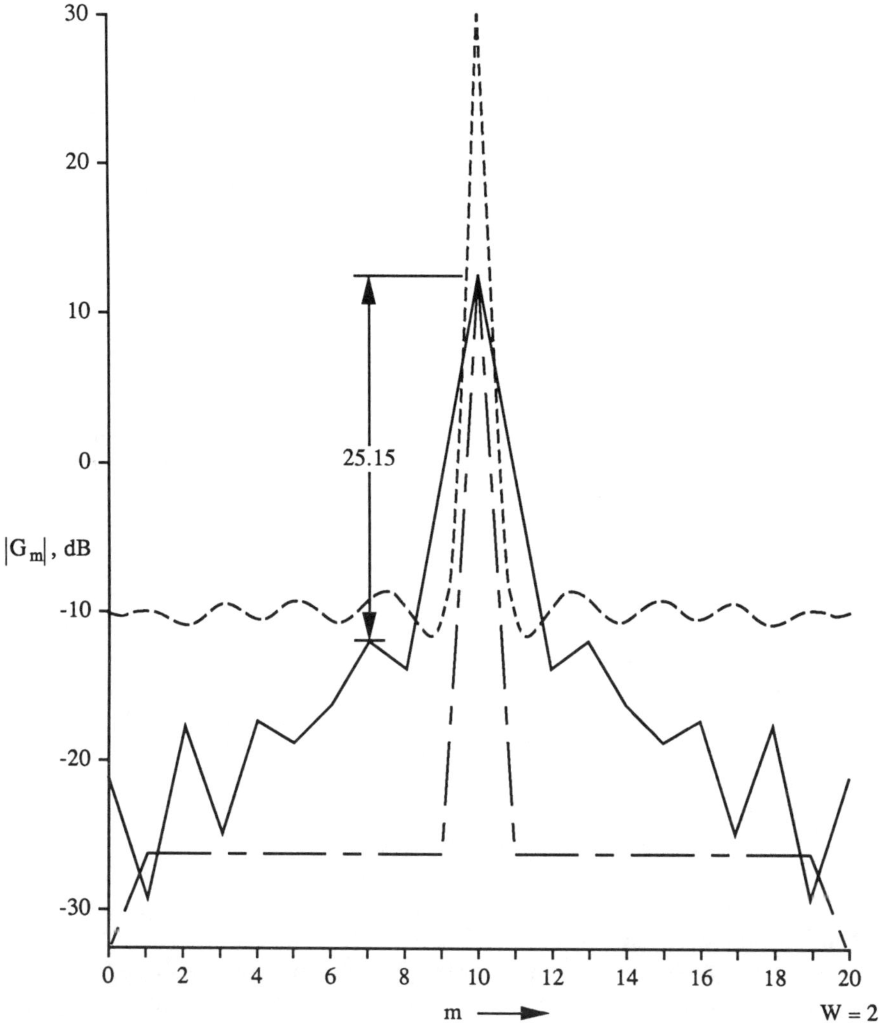

Figure 6.4 The ME spectrum $|G_m|^2$ of Eq. (6.35) versus frequency $f = m\Delta f' = 0.1m$. Each spectrum is plotted in dB relative to its own average power. Dashed line: the Gibbs-Burg spectrum of Eq. (5.19) with 10 percent white noise added. From [6.2] Fig. 10. Broken line: the N = 19, M = 20 noiseless spectrum Eqs. (6.17) to (6.19) of Fig. 6.3. Solid line: the N = 10, M = 20 noiseless spectrum of Fig. 6.3.

Burg curve is better than that of the N = 19 Boltzmann curve, which is better than that of the N = 10 curve.

If noise were present in the Boltzmann data processing it would start to corrupt seriously the ME amplitude spectrum when the average noise power becomes $\gtrsim$ noiseless spectral power -6 dB at any frequency of interest. According to this criterion 10 percent noise power would not corrupt the signal resolution at $f = W/2$ for either N = 19 or 10 but would seriously corrupt it at all other frequencies if N = 19 and for all frequencies outside the range $0.9 \leqslant f \leqslant 1.1$ if N = 10.

Our second example is the spectrum of a noiseless Ricker wavelet, popular in geophysical seismic work

$$y(t) = (-1 + 0.03t^2)\exp\left(\frac{-t^2}{\nu}\right), \quad \nu = \frac{200}{3} \tag{6.37}$$

This function is essentially zero for $t > 20$. If we sample at $\Delta t = 1$ and truncate the measurement at T = 20 then, N = 20 and $W = 1/(2\Delta t) = 0.5$. The Nyquist sampling rate is $\Delta f = 1/(2T) = 0.025$. The ME spectrum obtained at intervals $\Delta f' = 0.02$ for $M = W/\Delta f' = 25$ is plotted in Fig. 6.5 by the broken dotted line, up to $f = 0.09$. Beyond this the ME spectrum is negligible, as expected, up to W = 0.5. This prediction is rather unexpected, considering the coarse time resolution of $\Delta t = 1$. Our algorithm converged in about 6 iterations with an error of less than 2 percent in the ME-estimated value of each y_n at $t = n\Delta t$ satisfying Eq. (6.4) as well as the constraints embodied in the bracketed expression of Eq. (6.8) equal to zero.

Shown also in Fig. 6.5 are the MA (moving-average, all-zero), the ARMA (pole-zero), and the AR (autoregressive moving-average or Gibbs-Burg) spectra for comparison [6.3].

To assess the potential of our Boltzmann method of computing an ME wavenumber spectrum from spatial array data we studied the angular resolution capability of a linear array of 10 receivers spaced $\lambda/2$ apart, as shown in Fig. 6.6. The receiving elements are in and perpendicular to the $y - z$ plane and the sources are in that plane, radiating coherent electric field parallel to the receivers. The data processing constraint equations are Eq. (6.14) for N = 10 receivers and the algorithm solved Eqs. (6.26) to (6.33) self-consistently. We superresolved two distant coherent sources less than λ/L radians apart, which is the classical resolution limit for two nearly broadside sources. As explained by Gabriel [6.5], such superresolution is very difficult to obtain unless one resorts to artificial methods of data processing such as aperture receiver modulation.

Figure 6.6 shows the ME resolution of two noiseless point sources indicated by arrows, separated by $\theta = 6.3° = 0.50\ \lambda/L$, both when they are nearly broadside ($\theta \cong 90°$) and also nearly 45° from broadside. Both sources have amplitudes $G_1 = G_2 = 1\underline{/0°}$. For each mean direction of incidence we *assumed* the sources had been localized by other measurements to a scan

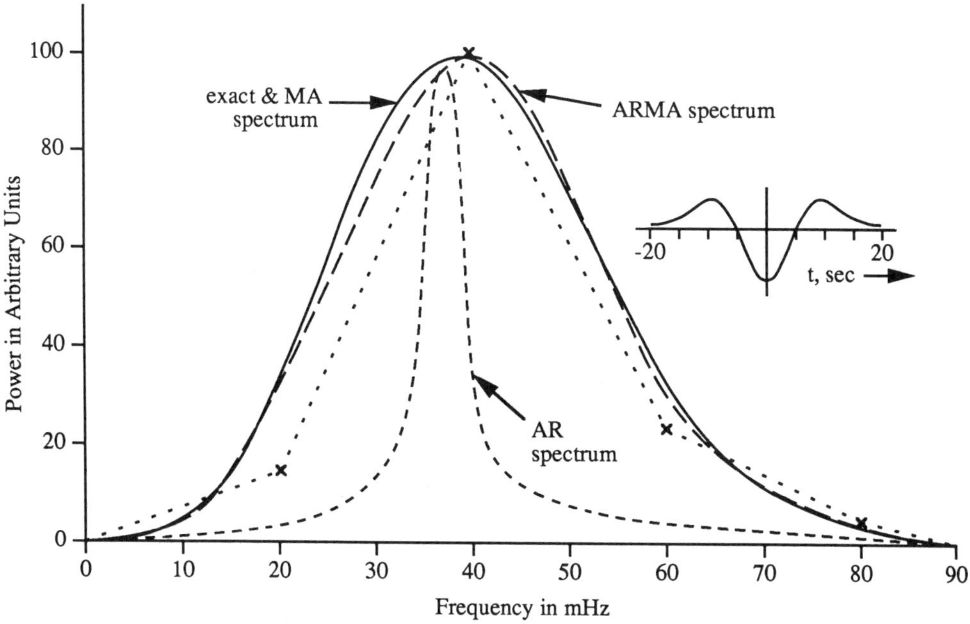

Figure 6.5 The relative Ricker wavelet power spectrum of Eq. (6.37) estimated by various methods. Our ME spectrum for M = 25 spectral points, superresolved with discrimination $\Delta f' = W/M = 0.5/25 = 0.02$ by N = 20 temporal points ($\Delta f = 0.025$) is shown by the dotted broken curve. The AR (autoregressive or all-pole) Gibbs-Burg spectrum Eq. (5.19), the MA (moving-average or all-zero) spectrum, and the ARMA (autoregressive moving-average or pole-zero) spectrum are reproduced from Fig. 3 of [6.3]. [©1978, IEEE]

region $\pm 12°$ on either side of one of the sources. Then, we assumed 21 *possible* sources at intervals of 1.2° within this region. In other words, N = 10 receivers resolved M = 21 possible source directions and the associated sources.

The solid broken line in Fig. 6.6 shows what one would reasonably call superresolution of the two nearly broadside sources in angle, in relative amplitude, and in phase as well—even though *we put no constraints on the phase behavior*. The curve indicates we can expect angular superresolution better than 0.50 λ/L at broadside. Computations for the two sources separated by 0.25 λ/L at broadside did, in fact, yield relative peaks at the true source angles but their amplitudes differed by a factor of 1.5. *Their phases were about as accurate as before!*

The dashed broken line of Fig. 6.6 shows the comparable superresolution in angle, amplitude, and phase for the two sources at nearly 45° to broadside and separated by 0.50 λ/L radians. This indicates sharper resolution compared to the classical limit than for broadside incidence because

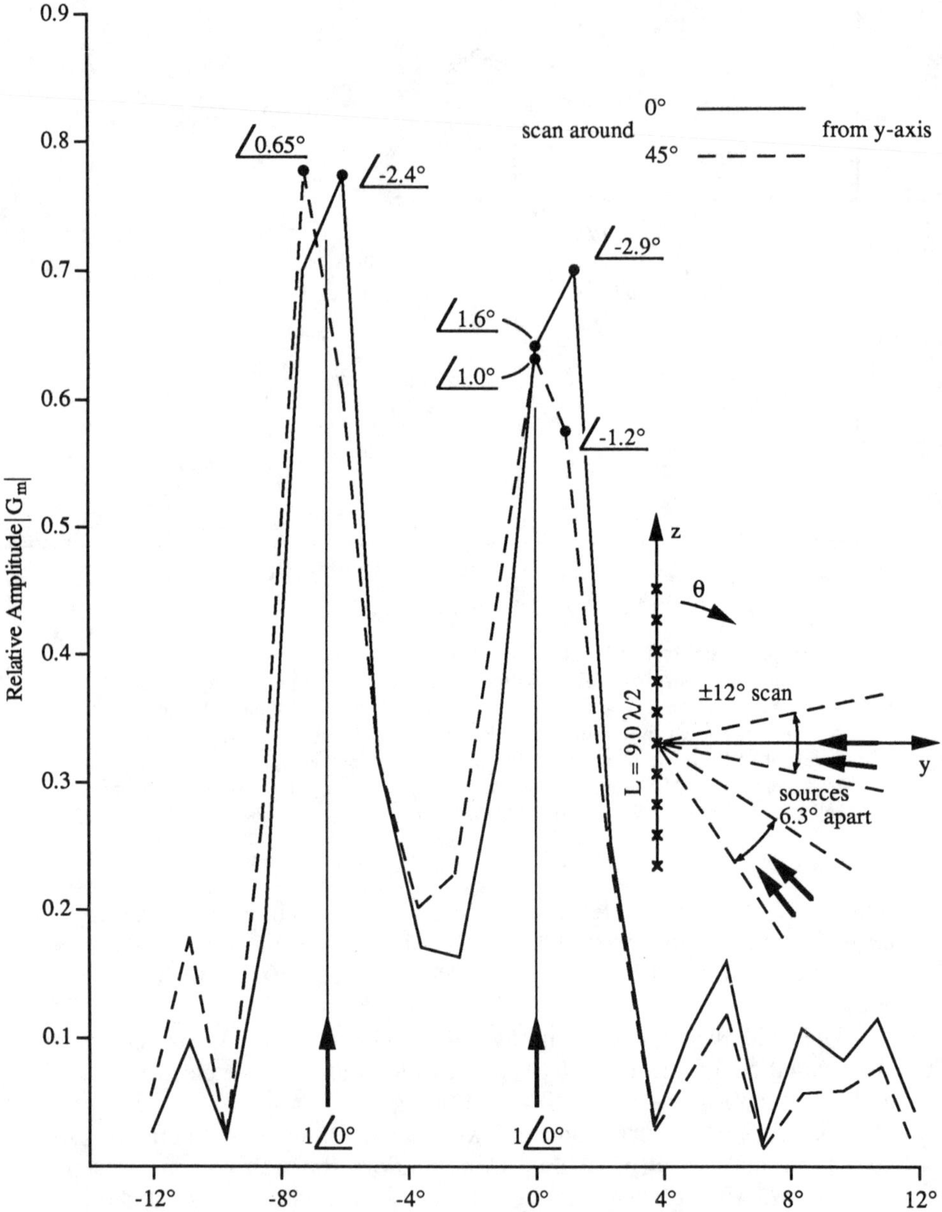

Figure 6.6 The relative Boltzmann wavenumber spectrum for the two coherent sources of 1/0° separated 6.3° spatially (0.50 λ/L), resolved by the 10-receiver array shown. The receivers resolved possible sources in 21 specified directions in each scan sector. From [6.4] Fig. 3.

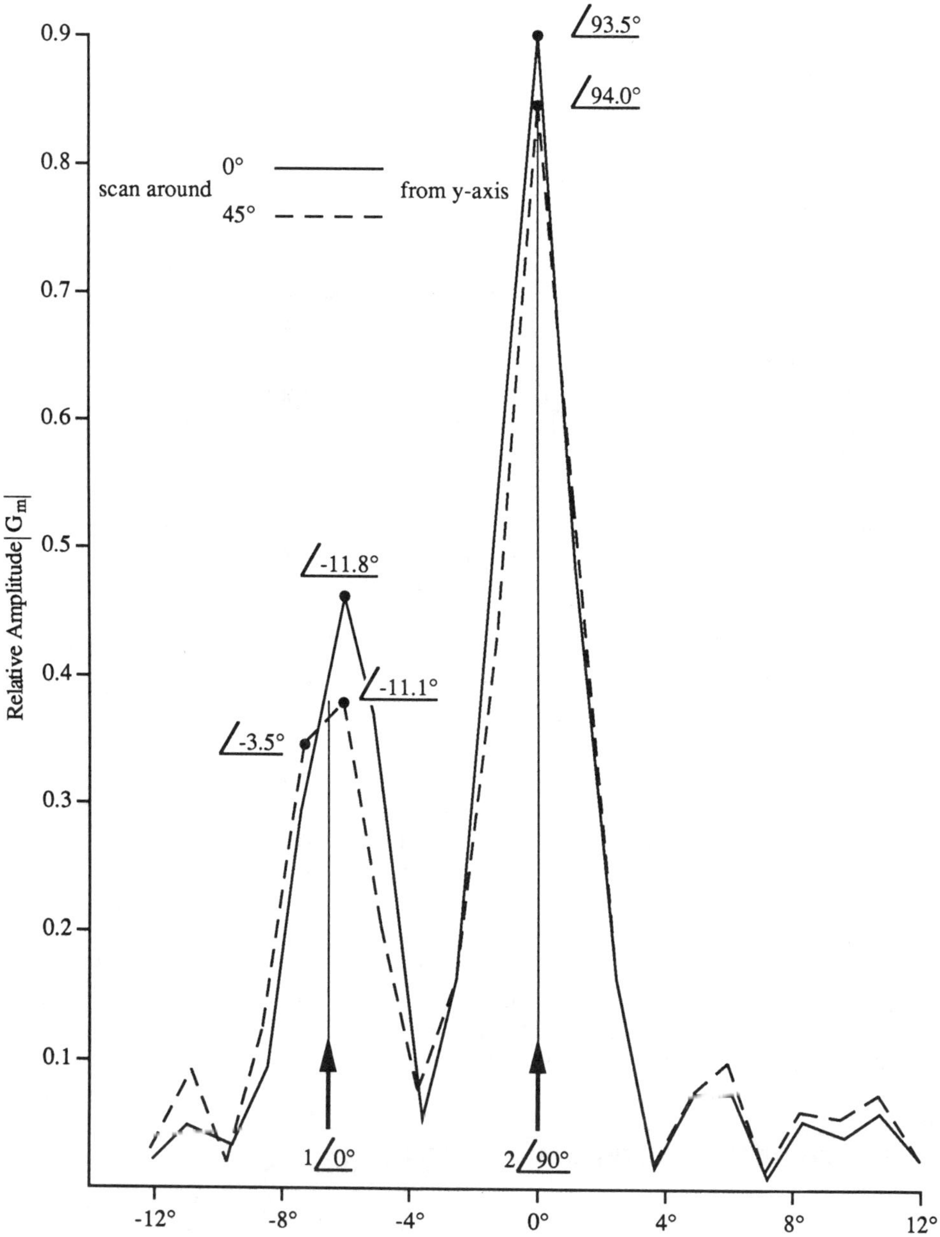

Figure 6.7 The relative Boltzmann wavenumber spectrum for the two coherent sources of $1\underline{/0°}$ and $2\underline{/90°}$ separated 6.3° spatially (0.50 λ/L), resolved by the 10-receiver array shown in Fig. 6.6. The receivers resolved possible sources in 21 directions in each scan sector. From [6.4] Fig. 3.

the classical limit for incidence 45° from broadside is measured by the array length perpendicular to this direction: $\lambda/(L \cos 45°) = 1.41\ \lambda/L$.

Lest the reader believe the accurate phase predictions were fortuitous for 0° phase sources, we show in Fig. 6.7 the ME superresolution for the same source locations as in Fig. 6.6 but with the strengths $1\underline{/0}$ and $2\underline{/90°}$, as indicated. The superresolution is comparable for both scan regions, in angle, magnitude, and phase. The predicted phase for the $1\underline{/0°}$ source was in error about 11°; for the $2\underline{/90°}$ source, about 4°.

These computational results were obtained on a CDC 7600 computer, of single-precision numbers with 14 significant decimal digits. Computers of shorter word size might be unable to invert Eq. (6.23) for the frequency spectrum or Eqs. (6.30) to (6.31) for the wavenumber spectrum and thereby solve for the $\{\beta_n\}$ from the $\{y_n\}$ or $\{Y_n\}$ data each iteration of the algorithm as it converges toward the ME solution. However, this algorithm and other similar ones for the Boltzmann ME spectrum, either frequency or wavenumber, are theoretically better behaved than the algorithm for the Boltzmann geotomography problem in Chap. 3. There the I (number of constraints) $\lesssim$ K (number of unknowns)/2 limit was imposed by the nature of the ME generalized inverse solution. Here, for the ME spectrum resolution, there is no such limit and the generalized-inverse basis of the solution *always exists* for any $N < M$—as we will demonstrate in the next section. Evidence of this general behavior is furnished by the single-frequency resolution for $N = 19$, $M = 20$ and by the Ricker wavelet resolution for $N = 20$, $M = 25$.

To conclude this section we observe that the Boltzmann ME algorithm of Sec. 6.5 has been written to resolve coherent temporal signals sampled at equal time intervals, or coherent radiating sources by a line of equally spaced receivers. The "equal time or spatial interval" is a prerequisite for the usual Gibbs-Burg ME data processing of Sec. 5.3. However, for Boltzmann ME data processing the temporal signal samples could be unequally spaced, and the antenna receivers could lie at arbitrary positions and the general analysis would apply. The full power of the Boltzmann method is applicable to the general problem of resolution of distant coherent radiators in azimuth ϕ as well as elevation θ in Fig. 6.6.

6.7 STABILITY, STATISTICAL UNCERTAINTY, AND NOISE

The fact was mentioned that we have resolved frequency spectra of coherent sources when N, the number of constraints or time samples, is close to M, the number of resolved frequencies (for $N = M$ there is only one solution and no need of the ME machinery). There is no limit like $N \simeq M/2$, analogous to the $1 \lesssim K/2$ limit imposed for geotomographic data processing in Chap. 3.

We now derive the form of the generalized inverse expression for wave-number spectrum **G** from Eqs. (6.26) to (6.28) and Eqs. (6.30) to (6.33). The corresponding matrix equations and generalized inverse expression for the frequency spectrum have the same properties. The generalized inverse expression for G will indicate the benign nature of the Boltzmann ME solution.

Equations (6.26), (6.27) in matrix form, with their row by column dimensions indicated, are

$$\mathbf{G}_{2M\times 1} = -(\mathbf{L}_\mathbf{D}^{-1})_{2M\times 2M}\mathbf{P}_{2M\times 2N}\boldsymbol{\beta}_{2N\times 1} \qquad N < M \tag{6.38}$$

where

$$\mathbf{G}_{2m-1,1} = G_{mR}, \qquad \mathbf{G}_{2m,1} = G_{mI}$$

$$(\mathbf{L}_\mathbf{D})_{2m-1,2m-1} = (\mathbf{L}_\mathbf{D})_{2m,2m} = -2L_m > 0, \quad (\mathbf{L}_\mathbf{D} \text{ is diagonal})$$

$$\mathbf{P}_{2m-1,2n-1} = \mathbf{P}_{2m,2n} = C_{mn}, \qquad \mathbf{P}_{2m-1,2n} = S_{mn}, \qquad \mathbf{P}_{2m,2n-1} = -S_{mn}$$

$$\boldsymbol{\beta}_{2n-1,1} = \beta_{nR}, \qquad \boldsymbol{\beta}_{2n,1} = \beta_{nI} \tag{6.39}$$

Equations (6.30), (6.31) in matrix form are

$$\mathbf{Q}_{2N\times 2N}\boldsymbol{\beta}_{2N\times 1} = \mathbf{Y}_{2N\times 1}, \qquad \text{or} \qquad \boldsymbol{\beta} = \mathbf{Q}^{-1}\mathbf{Y} \tag{6.40}$$

where

$$\mathbf{Q}_{2n'-1,2n-1} = \mathbf{Q}_{2n',2n} = F_{n'n}, \qquad \mathbf{Q}_{2n'-1,2n} = G_{n'n}$$

$$\mathbf{Q}_{2n',2n-1} = -G_{n'n}, \qquad \mathbf{Y}_{2n-1,1} = Y_{nR}, \ \mathbf{Y}_{2n,1} = Y_{nI} \tag{6.41}$$

Now **Q** can be written in terms of **P** by Eqs. (6.32), (6.33) as

$$\mathbf{Q}_{2N\times 2N} = -(\mathbf{P}^t)_{2N\times 2M}(\mathbf{L}_\mathbf{D}^{-1})_{2M\times 2M}\mathbf{P}_{2M\times 2N} \tag{6.42}$$

$\mathbf{P}^t$ being the transpose of **P**. We can now express the spectrum matrix **G** of Eq. (6.38) in terms of data matrix **Y** of Eq. (6.40), with $\mathbf{Q}^{-1}$ obtained from Eq. (6.42):

$$\mathbf{G} = \mathbf{L}_\mathbf{D}^{-1}\mathbf{P}\cdot(\mathbf{P}^t\mathbf{L}_\mathbf{D}^{-1}\mathbf{P})^{-1}\cdot\mathbf{Y} = \mathbf{A}^-\mathbf{Y} \tag{6.43}$$

$\mathbf{A}^-$ is the generalized inverse matrix, analogous to that of Eq. (3.33).

Equation (6.43) corresponds to the generalized inverse solution Eq. (3.36), which requires the following matrix rank condition for a meaningful $\mathbf{A}^-$

$$\text{Rank}\,(\mathbf{P}^t\mathbf{L}_\mathbf{D}^{-1}\mathbf{P}) = \text{Rank}\,(\mathbf{P}^t) \tag{6.44}$$

According to remarks following Eq. (3.37) this reduces to

$$\text{Rank}\,(\mathbf{P}^t\mathbf{L}_\mathbf{D}^{-1/2}) = \text{Rank}\,(\mathbf{P}^t) \tag{6.45}$$

Now $\mathbf{L}_{\mathbf{D}}^{-1/2}$ is a diagonal matrix operating on the columns of $\mathbf{P}^{t}$: column 2m-1 is multiplied by $1/\sqrt{-2L_m}$ for example. These *column* factors do not affect the number of rows of $\mathbf{P}^{t}\mathbf{L}_{\mathbf{D}}^{-1/2}$ which can be expressed as linear combinations of the remaining rows. Therefore, the rank of $\mathbf{P}^{t}\mathbf{L}_{\mathbf{D}}^{-1/2}$ is the same as that of $\mathbf{P}^{t}$, Eq. (6.45) is satisfied for any $N < M$ and *the generalized matrix solution Eq. (6.43) exists*! However, it must be iterated to a self-consistent solution.

In the process of iterating the matrix equations to a solution the matrix $\mathbf{Q}$ in Eqs. (6.42) and (6.43) might approach singular behavior, in which case the $\mathbf{G}$-values would be physically too large. Readjustment of their values and the $\mathbf{L}_{\mathbf{D}}$ values would alleviate this situation. As $N \rightarrow (M\text{-}1)$ algorithmic convergence can become a problem unless the iteration procedure is improved.

Unfortunately, there is no extension of the Darwin-Fowler analysis of Sec. 3.9 for statistical uncertainty of a spectral component G_{mR} or G_{mI} about its ME-value, within the ensemble of possible objects all satisfying the measurement constraints. That analysis is appropriate only for distributions of real positive (or negative) object parameters.

We have observed computationally that receiver noise which effectively alters even slightly the magnitude and/or phase of the received signals can seriously corrupt an ME solution for the wavenumber spectrum. It is vital to remove such noise by subtraction and/or correlation techniques as much as possible in order to recover the correct magnitude and phase of each receiver signal.

We can account for coherent object noise in the measurements, shown in Fig. 3.8, by a perturbation formula from Eq. (6.43) for the spectral noise δG induced by measurement noise δY. This would represent object noise in first approximation only because $\mathbf{L}_{\mathbf{D}}$ is itself a function of $\mathbf{G}$ in Eqs. (6.20) and (6.28). Allowing linear changes in the spectrum (object) about its ME value we have

$$\delta\mathbf{G} = -(\mathbf{U} - \mathbf{A}^{-}\mathbf{A})\mathbf{L}_{\mathbf{D}}^{-1}\,\delta\mathbf{L}_{\mathbf{D}}\mathbf{G} + \mathbf{A}^{-}\delta\mathbf{Y} \tag{6.46a}$$

where $\mathbf{A} = \mathbf{P}^{t}$ and Eq. (6.14) is $\mathbf{A}\,\mathbf{G} = \mathbf{Y}$. Note that the perturbed solution Eq. (6.46a) satisfies $\mathbf{A}\,\delta\mathbf{G} = \delta\mathbf{Y}$. If we neglect the first term in Eq. (6.46a) we have

$$(\delta\mathbf{G})_i = \sum_j (\mathbf{A}^{-})_{ij}(\delta\mathbf{Y})_j \tag{6.46b}$$

with a variance of

$$\langle (\delta\mathbf{G})_i^2 \rangle \approx \sum_{jk} (\mathbf{A}^{-})_{ij}\,(\mathbf{A}^{-})_{ik}\,\langle (\delta\mathbf{Y})_j(\delta\mathbf{Y})_k \rangle \tag{6.47}$$

This is analogous to Eq. (3.58), in which the last or statistical uncertainty term is neglected.

REFERENCES

[6.1] E. Rietsch, The maximum entropy approach to inverse problems, *Jour. Geophys., 42*, 1977, pp. 489–506.

[6.2] J.P. Burg, Maximum entropy spectrum analysis, In *Modern Spectrum Analysis*, D.C. Childers (ed.). IEEE Press, 1978, pp. 34–41.

[6.3] P.R. Gutowski, E.A. Robinson, and S. Treitel, Spectral estimation: Fact or fiction, IEEE Trans. Geosci. Elect., *GE-16*, April, 1978, pp. 80–84. Reprinted In *Modern Spectrum Analysis*, D.C. Childers (ed.). IEEE Press, 1978, pp. 49–53.

[6.4] R.M. Bevensee, Maximum entropy methods in electromagnetic/geophysical/ ultrasonic imaging, In *Inverse Methods in Electromagnetic Imaging* (Proceedings of the NATO Advanced Research Workshop, W. Germany, Sept. 1983), W-M Boerner, H. Brand, L.A. Cram, D.T. Gjessing, G. Schwierz, and M. Vogel (eds.). Part 1, pp. 375–395, D. Reidel, Dordrecht, 1984.

[6.5] W.F. Gabriel, Spectral analysis and adaptive array superresolution techniques, Proc. IEEE, *68*, June, 1980. pp. 654–666.

7

Maximum Entropy Photon Image Restoration

7.1 INTRODUCTION

This chapter discusses the inverse problem of reconstructing a brightness pattern of an object subdivided into square pixels, as obtained by either the measured image brightness pattern or the measured image wavenumber spectrum. We review two-dimensional image work of other researchers; three-dimensional tomographic problems could be solved by extending these methods. The image plane, also subdivided into square pixels, is taken parallel to the object plane and measurements are made within a narrow frequency range $\Delta\nu$.

In Sec. 5.6 we applied the Burg method to construct an object wavenumber spectrum $S_\omega(k_x)$ from an ME-derived image power spectrum $\mathcal{P}(\rho_x\omega)$—Eq. (5.78). Such a power spectrum for $\rho_x = 0$ is also called the *image* brightness pattern. An *object* brightness pattern refers to a spatial distribution of sources radiating within a narrow frequency range, where the power intensity from pixel k is proportional to the number of photons $n(k)$ radiated during the measurement time.

In Sec. 7.2 the objective function F is the entropy of the relative object brightness plus the independent constraints imposed by a total relative object brightness of 1, measurements, and the known direct transmission factors from the various object pixels to image pixels. A parameter β is set numerically to constrain the "noise" in the ME solution to be close to the measurement noise. $(+ \sum \log p_k)$ is employed for object entropy, k denoting the kth pixel, where $p_k = \langle n(k)\rangle / \sum_k \langle n(k)\rangle$ measures average photon intensity or brightness in pixel k and $\langle n(k)\rangle$ is the expected number of photons emitted from that pixel during measurement time T. The reason for selecting this entropy form instead of the Shannon $-\sum p_k \log p_k$ involves the number of degrees of photon freedom for each pixel, relative to $\langle n(k)\rangle$ and is explained in Sec. 7.3.

In Sec. 7.3 we follow Kikuchi and Soffer and derive, from first principles, the ME distribution for $n(k)$, a possible number of photons, emitted from pixel k, during measurement time T, in terms of the expected number $\langle n(k)\rangle$ regarded as an observable. The maximum entropy turns out to be proportional either to $\log(\langle n(k)\rangle)$ or to $-\langle n(k)\rangle \log(\langle n(k)\rangle/z)$, depending on whether the number of photon degrees of freedom z is such that $z \ll \langle n(k)\rangle$ or $z \gg \langle n(k)\rangle$, respectively. z consists of a temporal factor times a spatial factor and should be carefully estimated for every photon image restoration problem in order to select the correct entropy expression for each pixel in the objective function.

In Sec. 7.4 we review the analysis of Wernecke and D'Addario, who derived ME object spatial brightness patterns from their measured image brightness patterns in wavenumber space. Wavenumber brightness is the Fourier transform of the spatial brightness pattern. A noise constraint is expressed as an upperbound on the sum of the $|\text{ measured } - \text{ ME } |^2$ errors, normalized to their known standard deviations of the noise. An example from radio astronomy shows the superiority of their ME method over direct Fourier transformation to resolve the spatial brightness pattern of a radio star.

7.2 RECONSTRUCTION OF OBJECT BRIGHTNESS FROM IMAGE BRIGHTNESS IN REAL SPACE

The underdetermined inverse problem of ascertaining the most probable ME coherent, partially coherent, or incoherent relative brightness p_k distribution of an object from its degraded photon image pattern b_i is depicted in Fig. 7.1. The entropy in an (unknown) relative brightness pattern $\{p_k\}$ over K object pixels radiating *independently* is taken to be

$$h = \sum_{k=1}^{K} \log p_k < 0, \qquad p_k = \frac{\langle n(k)\rangle}{\sum_j \langle n(j)\rangle} \tag{7.1}$$

This entropy has only one maximum relative to all the $\langle n(k)\rangle$, $\sum_k \langle n(k)\rangle$ being unconstrained, because all the $\partial^2 h/\partial\langle n(k)\rangle^2$ are <0 at every $\{\langle n(k)\rangle\}$ point for $K > 1$, regardless of the specific constraints. This form for h corresponds to case (ii) of Sec. 7.3, when the number of photon degrees of freedom z is $\ll$ each $\langle n(k)\rangle$.

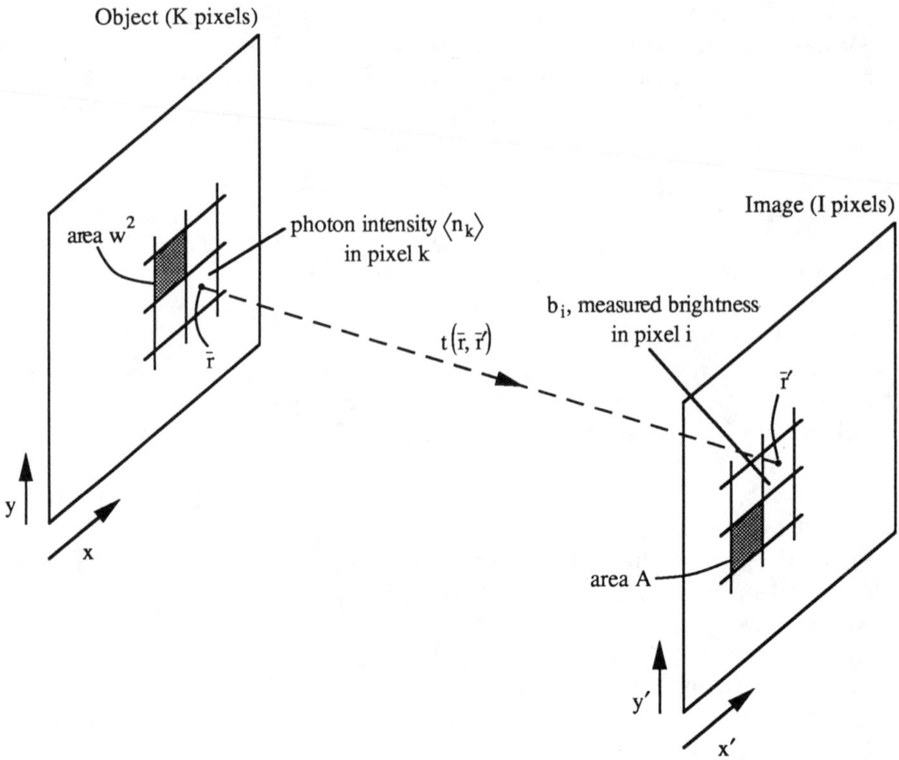

Figure 7.1 Photon imaging of an object by direct transmission of photons from object to image. $p_k = \langle n_k \rangle / \sum_k \langle n_k \rangle$ is the relative brightness of object pixel k of area w^2 at $\bar{r}$. $t(\bar{r}, \bar{r}')$ is the direct transmission factor from $\bar{r}$ to image pixel i at $\bar{r}'$. b_i is the measured brightness in that image pixel of area A.

The constraint

$$\sum p_k = 1 \tag{7.2}$$

will be imposed in the ME-Boltzmann solution.

Following [7.1], we employ the direct transmission factor

$$t(\bar{r}, \bar{r}') = \left[\frac{\sin(x - x')}{x - x'} \frac{\sin(y - y')}{y - y'} \right]^2 \tag{7.3}$$

or "point-spread" function, so-called because it describes the spread of radiation from a source at position $\bar{r}$ to various $\bar{r}'$ on the image. In Eq. (7.3), distances are normalized to wavenumber $k = 2\pi/\lambda$. Other point-spread functions are appropriate in other situations.

The I ($<$ K) constraints on the object relative brightness spatial distribution on K pixels imposed by the b_i measurements in Fig. 7.1. are

$$\sum_{k=1}^{K} p_k(\bar{r}_k)t(\bar{r}_k, \bar{r}_i') - b_i(\bar{r}_i') = 0, \qquad 1 \le i \le I < K \tag{7.4}$$

Rather than impose an independent constraint like Eq. (7.4) on each of the I image pixels and introduce I Lagrange multipliers we follow Kikuchi and Soffer [7.1] and impose a single constraint on the total "noise energy," *defined* as the sum of all the squared deviations from zero in Eq. (7.4). This energy E is a function of the $\{p_k\}$,

$$E(\{p_k\}) = \sum_{i=1}^{I} \left[\sum_{k=1}^{K} p_k(\bar{r}_k)t(\bar{r}_k, \bar{r}_i') - b_i(\bar{r}_i') \right]^2 \tag{7.5}$$

The objective function F, which Kikuchi and Soffer liken to the free energy of a thermodynamic system, is

$$F = h \text{ [of Eq. (7.1)]} + \beta E(\{p_k\}) + \gamma \left(\sum p_k - 1\right) \tag{7.6}$$

β and λ being the Lagrange multipliers.

There are several ways of deriving an ME $\{p_k\}$ from this F. We could set a variation of $\delta F = 0$ for each δp_k and thereby obtain K equations for the $\{p_k\}$ involving β and γ, then solve for all of these self-consistently by an iterative scheme, so as to enforce Eq. (7.2) exactly and $E(\{p_k\})$ within an interval. This interval would be set by the known standard deviation σ of the image noise (see Fig. 1.8) in each p_k independently of all the other p_j,

$$E(\{p_k\}) \le I\sigma^2 \tag{7.7}$$

Such a solution, according to the prescription of Eq. (1.33b) in Sec. 1.4, might be difficult to find if the image noise level is high or if an unrealistically low value of σ^2 is assumed. Indeed, if σ^2 is set to zero no value of β will constrain E to zero. Instead, we will pursue the simpler procedure of fixing a value for β and follow the previous iterative method but enforce only the one constraint Eq. (7.2) to obtain γ.

No preselected value of β will constrain E to zero. A small β will generate a rather uniform object with little detail. As β increases the ME object becomes sharper and more detail appears. Image noise would blur it. Too large a β will create a very "touchy" numerical iterative solution, of either erratic variation about an "averaged" solution or slow and erratic convergence to the final ME solution. Such behavior will occur whether or not image noise is present. Either type of behavior can achieve superresolution in the sense that the object is recognizable even though resolved with image pixel areas smaller than the resolution limit set by the image pixel (detector) area A. Superresolution can even occur when the image is diffraction-limited, meaning that the point-spread functions of more than one object pixel overlap appreciably on the same image pixel.

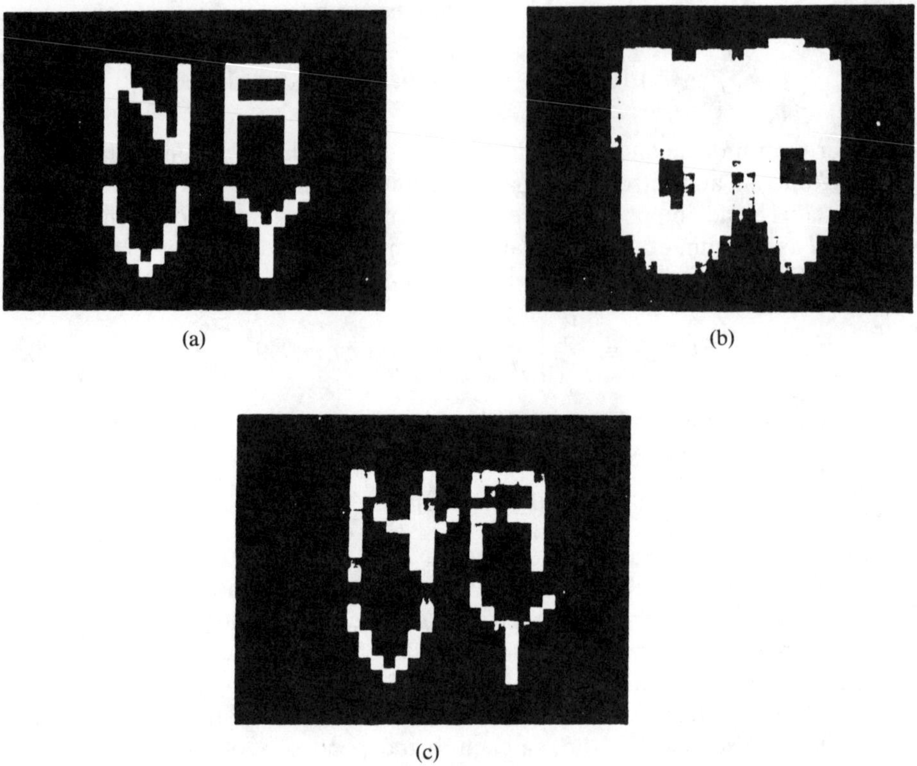

Figure 7.2 ME superresolution of a binary object (pixel brightness either 0 or 1) in Fig. 7.1. Transmission factor t of Eq. (7.5) is given by Eq. (7.3): (a) alphabetic test object on a 20×20 grid; (b) its diffraction limited image; (c) the ME object estimate from the objective function Eq. (7.6), with $\beta = 10^7$. The gray areas in (c) are uncertain because of pattern fluctuations in successive iterations of the ME procedure. From [7.1] Fig. 14.

If one has some a priori information to aid the object identification it could help to resolve the object parameters of interest in a sequence of ME solutions generated by a sequence of increasing β in Eq. (7.6), to a limit imposed by the actual S/N ratio. Fig. 7.2 shows an impressive synthetic-data example of ME superresolution of a binary object (pixels either black with intensity 0 or white with intensity 1), for a final value of $\beta = 10^7$. There is no image noise, only noise energy. Convergence to this ME solution was slow and erratic, with large object fluctuations between consecutive iterations, until the final reconstructed object was obtained. Part of the discrepancy between the true object in Fig. 7.2(a) and the ME object of Fig. 7.2(c) is due to an unconstrained $E > 0$ in Eq. (7.5). The gray areas in the ME object are more uncertain than the others. Even if E were focused to zero or to a minimal value by introducing more Lagrange multipliers the reconstructed object would not necessarily be significantly better than shown in Fig. 7.2(c).

Clearly, if one knew the object in Fig. 7.2 consisted of English letters, one would be quite certain of the true object.

7.3 STATISTICS OF PHOTON EMISSION, TRANSMISSION, AND DETECTION [7.2]

Our objective in this section is to derive the general entropy expression $h(k)$ for the brightness or power density radiated from a typical object pixel k, within a narrow frequency range $\Delta\nu$, in terms of the expected number of photons $\langle n(k)\rangle$ emitted during measurement interval T. $\langle n(k)\rangle$ is obtained by averaging over the ME probability distribution of possible $n(k)$. $\langle n(k)\rangle$ is an estimate of the average of the $n(k)$ from a large number of measurements, each of duration T.

Knowledge of $h(k)$ for a typical pixel k enables us to write the total object entropy $h_T = \sum_k h(k)$, treating the pixels as independent radiators. h_T is then maximized, subject to constraints, in an objective function for a larger problem of photon object resolution from its image.

In the remainder of this section we suppress the index k for the kth pixel, with the understanding that one pixel is under discussion.

Since the Kikuchi-Soffer procedure for obtaining entropy $h(k) = h$ in terms of $\langle n(k)\rangle = \langle n\rangle$ is rather involved we summarize the steps in the reasoning as follows:

1. From the coherence volume V_{coh} of the photons, coherence bandwidth $\Delta\nu$, and detector area A we obtain the number of degrees of freedom z for any number n of photons emitted during measurement time T.
2. The number of equiprobable microstates in a photon macrostate n is $q(n)$, given by the Bose-Einstein statistics of Sec. 1.3.3.
3. $M \gg \gg 1$ possible emission states are assumed for detection time T, and Mf_n of them each have n photons.
4. For the ensemble of states $\{Mf_n\}$ there are $\prod_n q(n)^{Mf_n} = Q$ distinct microstates.
5. There are

$$\Omega = \frac{M!}{\prod_n (Mf_n)!}$$

distinguishable arrangements of the M emission states.
6. Hence, the aggregate of distinct microstates in all the M possible emission states is C, the product of Q and Ω.

Maximization of H = log(C) subject to the constraint that $\sum nf_n =$ observable $\langle n \rangle$ yields the ME $\{f_n\}$ which were defined in step 3. The general ME expression for h = H/M is Eq. (7.21), in terms of $\langle n \rangle$ alone. This admits of a simpler form, effectively $\log(\langle n \rangle)$ or $-\langle n \rangle \log(\langle n \rangle)$, depending on whether $z \ll \langle n \rangle$ or $\langle n \rangle \ll z$, respectively. For the latter case the $\{f_n^{\text{ME}}\}$ are given by the Poisson distribution

$$f_n^{\text{ME}} = e^{-\langle n \rangle} \frac{\langle n \rangle^n}{n!} \qquad (\langle n \rangle \ll z) \tag{7.8}$$

We now furnish the details of each of these steps in succession, according to the Boltzmann method of counting microstates and macrostates.

7.3.1 Photon Coherence Volume, V_{coh}, and Degrees of Freedom, z

Consider in Fig. 7.3 radiation from a pixel of area w^2 in the object plane a distance R from the detector of area A parallel to the object plane. V_{coh} is the volume denoted by $l\sigma$ within which we can define one photon within the narrow frequency range $\nu \pm \Delta\nu/2$. $\Delta\nu$ could be the width of a spectral line of emission from the object, in which case it is $1/\tau$, τ being the coherence time of the emission process. Or $\Delta\nu$ could be the frequency discrimination 1/T imposed by the Nyquist sampling rule in Sec. 5.2. For either case, we designate $\Delta\nu = 1/\tau$; $c\tau = l$ is the coherence distance for one photon in Fig.

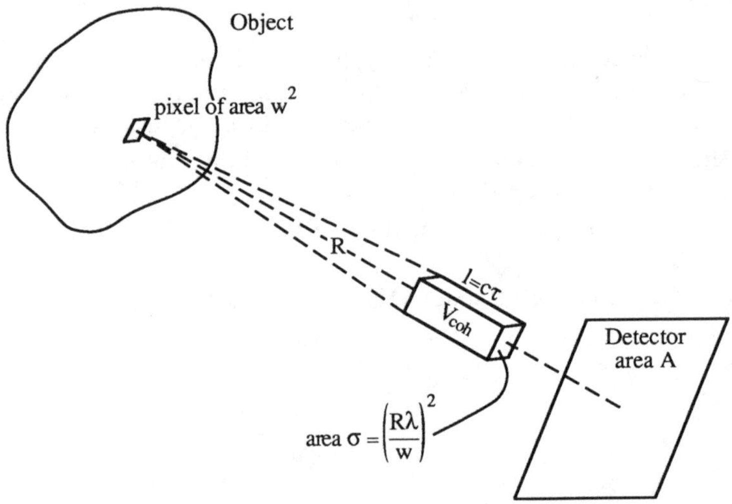

Figure 7.3 Photon radiation from an object pixel of area w^2 in the object plane to the detector of area A shown in the image plane. Object and image planes are parallel. Coherence volume $V_{coh} = c\tau\sigma$, given by Eq. (7.10).

7.3. The minimum area σ for distinguishing one photon in the far-field is given by the Zernecke-van Cittert Theorem [7.3].

$$\sigma = \left(\frac{R\lambda}{w}\right)^2 \qquad \lambda = \text{wavelength at } \nu \tag{7.9}$$

The square root of this is the classical angular resolution limit $\sqrt{\sigma}/R = \lambda/w$ for a distributed antenna of length w, mentioned in Sec. 6.6. Coherence volume V_{coh} is, therefore,

$$V_{coh} = l\sigma = c\tau\left(\frac{R\lambda}{w}\right)^2, \qquad \tau = \frac{1}{\Delta\nu} \tag{7.10}$$

The number of degrees of freedom, z, in measurement time T for the detector of area A is the number of coherence volumes in phase space volume cTA. Hence,

$$z = \frac{c\tau A}{V_{coh}} = z_t z_s = \left(\frac{T}{\tau}\right)\left(\frac{Aw^2}{R^2\lambda^2}\right) \tag{7.11}$$

z consists of two factors, a temporal one z_t and a spatial one z_s. For ordinary data processing in the time and frequency domains $\tau = 1/\Delta\nu = T$ according to the Nyquist sampling relation of Sec. 5.2 and so $z_t = 1$. For superresolution in frequency, $\Delta\nu < 1/T$, $z_t < 1$ and it is rounded up to 1 for the calculation of z. For long measurement times and τ fixed by an emission process, $T > \tau$ and so $z_t > 1$. Similar remarks apply to z_s in the spatial domain: it can be large or small but if <1 it is rounded up to 1 for calculating z. Therefore, z is always ≥ 1.

7.3.2 The Number $q(n)$ of Equiprobable Microstates in Photon Macrostate, n

In Sec. 1.3.3, Eq. (1.22) tells us the number of equiprobable photon microstates in the N_k macrostate of Fig. 1.6, where n of Eq. (1.23) is the number of degrees of freedom for a unit spatial volume $\Delta x \Delta y \Delta z = 1$. We now want to count the number of degrees of freedom z in the $\Delta p_x \Delta p_y \Delta p_z$ volume of photons radiated from area w^2 to the detector and falling within the "measurement" volume cTA.

Let the z axis of momentum p space lie in direction R. Then, $p_z \cong p = h\nu/c$. (Here h is Planck's constant.) The Δp_x range at any point on the detector is nearly pw/R, as it is for the Δp_y range. $\Delta p_z = h\Delta\nu/c$. Then, degrees of freedom n of Eq. (1.27) is

$$2\left(\frac{\nu}{c}\frac{w}{R}\right)^2 \frac{\Delta\nu}{c}$$

Upon mulitplying this by cTA and substituting $\nu = c/\lambda$, with $\Delta\nu = 1/\tau$, we obtain a second estimate, z_2, for z as

$$z_2 = 2\left(\frac{T}{\tau}\right)\left(\frac{Aw^2}{R^2\lambda^2}\right) \tag{7.12}$$

This is larger than Eq. (7.11) by factor 2 because Eq. (7.12) counts both possible states of photon polarization.

However, we shall continue to use Eq. (7.11) for the number of degrees of freedom for one polarization within the measurement volume cTA.

Now we can count the number of photon microstates $q(n)$ in a macrostate of n photons within volume cTA. It is given by Eq. (1.22) with z replacing n and the new n replacing N_k. Hence,

$$q(n) = \frac{(z + n - 1)!}{(z - 1)!\, n!} \tag{7.13}$$

7.3.3 Description of the M $\gg\gg$ 1 Possible Emission States

This is the crucial step in computing the entropy of a pixel in terms of the expected number of photons $\langle n \rangle$ reaching the detector in time T. Some of the possible emission states are shown in Fig. 7.4; each state pictures a sequence of photons arriving at various times during the interval T. Note, we say nothing about when the various photons are emitted in each state.

Group 3 in Fig. 7.4 shows a number Mf_3 of states, each with 3 photons. Group n has Mf_n states, each with n photons. f_n is the fraction of all M states with n photons, and its ME value will be found by maximizing the entropy $H = \log(C)$, C being the total number of distinct microstates among the ensemble of possible states in Fig. 7.4.

7.3.4 The Number of Microstates in the Ensemble $\{Mf_n\}$ of Emission States

There are Mf_n distinct states in Fig. 7.4 with n photons each. Each of these states has $q(n)$ distinct microstates according to Eq. (7.13). Any microstate can occur for each of the Mf_n states. Therefore, there are

$$q(n)^{Mf_n}$$

distinct microstates possible in group n and

$$Q = \prod_n q(n)^{Mf_n} \tag{7.14}$$

distinct microstates in the entire ensemble of M emission states.

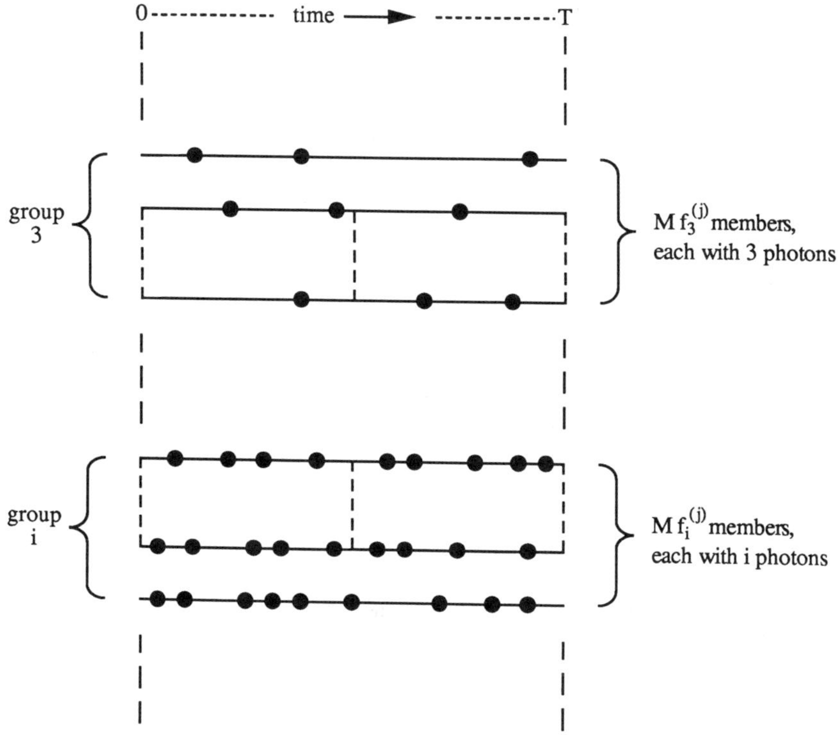

Figure 7.4 The ensemble of $M \gg \gg 1$ possible macrostates of photon emission from a pixel during measurement time T. A typical group i has Mf_i members, each with i photons emitted at various unspecified times. Each macrostate has a number $q(i)$ of distinct microstates, Eq. (7.13).

7.3.5 The Number of Distinguishable Arrangements of the M States

So far we have identified Q distinct microstates among the M emission states but we haven't said which emission state comes from what point on the pixel (recall we are discussing the photon statistics of one object pixel). If there are M points of emission we can assign any Mf_1 of them to emission state 1, Mf_2 points to emission state 2, and so on. The number of distinct arrangements or combinations of the M points among all the $\{Mf_n\}$ groups was derived in Sec. 1.3.1. Equation (1.3) gives the answer as

$$\Omega = \frac{M!}{\prod_m (Mf_m)!} \tag{7.15}$$

Each of these arrangements has Q microstates given by Eq. (7.14).

7.3.6 The Total Number of Distinct Microstates of Emission

This total number is the number Q of distinct microstates in any one arrangement of the emission states coming from M different points on the pixel, times the number Ω of distinct spatial arrangements of the $\{Mf_n\}$ states among the M points. Hence, the total number C of distinct microstates of emission from the pixel is

$$C = Q\Omega = \frac{M!}{\prod_m (Mf_m)!} \cdot \prod_n q(n)^{Mf_n} \tag{7.16}$$

This is the essential result of this section. Entropy H is log(C) and we shall calculate the $\{f_n\}$ to maximize it.

7.4 MAXIMIZATION OF THE ENTROPY OF A PIXEL RELATIVE TO ITS OBSERVABLE $\langle n \rangle$

It is now straightforward to maximize entropy H = log(C) of a pixel from Eq. (7.16), subject to the constraint that $\langle n \rangle = \sum_n nf_n$ is an observable which represents our state of knowledge about the emission process during measurement time T. This is a Gibbsian estimate of the expected number of photons emitted over the ensemble of M emission states.

From Eq. (7.16) we obtain, with the aid of Stirling's approximation Eq. (1.5),

$$h = H/M = \frac{\log(C)}{M} = -\sum_n f_n(\log f_n - 1) - 1 + \sum f_n \log[q(n)] \tag{7.17}$$

neglecting $(1 - \sum f_n)\log(M)$ because $\sum f_n$ will be constrained to 1.

The objective function for this pixel is

$$F = h + \mu\left(\sum nf_n - \langle n \rangle\right) + \beta\left(\sum f_n - 1\right) \tag{7.18}$$

That $\{f_n\}$ distribution which maximizes Eq. (7.18) with Lagrange multipliers μ and β chosen to constrain the parenthetical expressions to zero is

$$f_n^{ME} = \left(\frac{z}{\langle n \rangle + z}\right)^z \left(\frac{\langle n \rangle}{\langle n \rangle + z}\right)^n \frac{(n + z - 1)!}{n!(z - 1)!}, \tag{7.19}$$

negative binomial distribution

The variance of this is

$$\frac{\langle n \rangle(\langle n \rangle + z)}{z} \tag{7.20}$$

Equation (7.19) yields a maximum entropy from Eq. (7.17) of

$$h^{\mathrm{ME}} = (\langle n\rangle + z)\log(\langle n\rangle + z) - \langle n\rangle \log \langle n\rangle - z \log z \qquad (7.21)$$

in terms of $\langle n\rangle$ and z, as promised. $\langle n\rangle$ is proportional to expected brightness b of that object pixel.

Equation (7.21) is the grand result of Kikuchi and Soffer [7.2]—the general expression for the entropy of an object pixel in terms of the expected number $\langle n\rangle$ of photons emitted during measurement time T as well as the number of degrees of freedom z in Eq. (7.11), imposed by measurement conditions. Since all object pixels are assumed to radiate independently, we can reconstruct the brightness pattern of an object radiating within the narrow frequency range $\Delta\nu$ (which affects z) by maximizing the total object entropy $\sum_k h_k^{\mathrm{ME}}$ of all the pixels, using Eq. (7.21), relative to all the unknown $\langle n(k)\rangle$. The maximization is subject to the measurement constraints involving the transmission or point-spread factors between all object and image pixels. This procedure was followed in Sec. 7.2 and will be employed in Sec. 7.6.

7.5 SPECIAL CASES OF GENERAL ENTROPY EXPRESSION EQ. (7.21) FOR A PIXEL

7.5.1 $z = 1$

Now $q(n)$ of Eq. (7.13) reduces to unity so the entropy Eq. (7.17) reduces to the Shannon form $\sum f_n \log(f_n) \cdot h^{\mathrm{ME}}$ of Eq. (7.21) is

$$h^{\mathrm{ME}} = (\langle n\rangle + 1)\log(\langle n\rangle + 1) - \langle n\rangle \log \langle n\rangle \qquad (7.22)$$

If $\langle n\rangle \gg 1$ this reduces further to

$$h^{\mathrm{ME}} \approx 1 + \log \langle n\rangle, \qquad \text{``log B'' form, B denoting brightness} \qquad (7.23)$$

This is the classical wave-field limit of the entropy expression for Bose-Einstein statistics.

The "log(B)" form of entropy was employed by Burg [7.4] in his original derivation of the ME-frequency power spectrum of a time-limited function $f(t)$. See Sec. 5.3.

7.5.2 $\langle n\rangle \gg z > 1$

The general expression Eq. (7.21) reduces to

$$h^{\mathrm{ME}} = z \log \langle n\rangle - z(\log z - 1) \qquad (7.24)$$

Since z is a constant in a problem involving maximization of the entropy of all the pixels relative to their $\langle n\rangle$ and subject to object-image constraints, this h^{ME} again has the form of "log(B)."

Certain radio stars have enormous $\langle n\rangle/z$ ratios at megahertz frequencies.

7.5.3 $\langle n\rangle \ll z$

The negative binomial distribution Eq. (7.19) reduces to

$$f_n^{\mathrm{ME}} = \frac{e^{-\langle n\rangle}\langle n\rangle^n}{n!} \qquad \text{Poisson distribution} \tag{7.25}$$

and Eq. (7.21) becomes

$$h^{\mathrm{ME}} = \langle n\rangle \log z - \langle n\rangle(\log\langle n\rangle - 1) \tag{7.26}$$

Since $\partial h^{\mathrm{ME}}/\partial\langle n\rangle$ and $\partial^2 h^{\mathrm{ME}}/\partial\langle n\rangle^2$ have the same properties as for $-\langle n\rangle\log(\langle n\rangle)$, Eq. (7.26) is equivalent to the "$-$Blog(B)" form or Shannon form of entropy. This is the classical Maxwell-Boltzmann particle limit of the Bose-Einstein statistics.

Examples of bodies with $\langle n\rangle/z \ll 1$ include our sun in the visible spectrum and its planets in the visible and infrared portions of the spectrum.

In conclusion, the analysis of this section emphasizes the importance of employing the correct entropy function of the observables for an object radiating electromagnetic (photon) energy. In the next section we treat the resolution of radio astronomy objects for which $z \ll$ each $\langle n\rangle$ on each pixel and the "log(B)" form Eq. (7.24) is appropriate for entropy.

7.6 RECONSTRUCTION FROM IMAGE MEASUREMENTS IN WAVENUMBER SPACE

Wernecke and D'Addario [7.5] employed an ME-Boltzmann analysis to reconstruct radio astronomy maps of incoherent objects. They reconstructed spatial brightness or power spectrum patterns $\mathcal{P}_\omega(\bar{r})$ at a frequency ω by inverting the wavenumber spectral data $S_\omega(\bar{k})$ in the image plane. They employed the Fourier transform relation Eq. (5.69), extended to two dimensions [7.6],

$$S_\omega(\bar{k}) = \int dx \int dy\, \mathcal{P}_\omega(\bar{r})e^{-j\bar{k}\cdot\bar{r}} \tag{7.27}$$

Here $\bar{r} = \bar{a}_x x + \bar{a}_y y$ is the position vector in the object plane and $\bar{\mathrm{k}} = \bar{a}_x k_x + \bar{a}_y k_y$ is the wavenumber vector in the image plane. Equation (7.27) and its inverse Fourier transform are valid for reconstruction of a wide-sense, stationary, discrete-time random photon field.

In our notation spectrum S_ω is discretized as $S_m(k_{xm}, k_{ym})$ for $1 \le m \le \mathrm{M}$ and brightness $\mathcal{P}_\omega$ is discretized as $b_n(x_n, y_n)$ for $1 \le n \le \mathrm{N}$. Now the latter is underdetermined so we have M(image) < N(object). Equation (7.27) becomes

$$S_m(k_{xm}, k_{ym}) = A \sum_{n=1}^{N} b_n(x_n, y_n) \exp[-j(k_{xm}x_n + k_{ym}y_n)], \qquad 1 \le m \le M \tag{7.28}$$

A being the area of each object pixel.

The M measured data $\{S_m\}$ are assumed corrupted by independent zero-mean, random noise with standard deviations $\{\sigma_m\}$, so the noise imposes the gross constraint

$$\sum_{m}^{M} \frac{|\,S_m(\text{measured}) - S_m\,(k_{xm}, k_{ym})\,|^2}{\sigma_m^2} \le M \tag{7.29}$$

The entropy of object pixel i is taken to be $h_i = a\log(b_i)$ according to Sec. 7.5.2 for the number of degrees of freedom z (set by the measurement scheme) $\ll \langle n_i \rangle$ (proportional to b_i), the expected number of photons emitted during the measurement time. This inequality is typical of radio stars. The proportionality constants A and a do not affect the relative object-brightness patterns $\{b_n^{\text{ME}}\}$, which we obtain as follows.

The objective function F to extremize (maximize) in order to maximize the total entropy $h_\tau = \sum h_n$ of all the independent pixels subject to the single constraint Eq. (7.29) is

$$F = a\sum_{n}^{N} \log b_n - \beta\left\{ \sum_{m}^{M} \frac{1}{\sigma_m^2} \left| S_m\,(\text{measured}) - A\sum_{n}^{N} b_n(x_n, y_n) \exp[-j(k_{xm}x_n + k_{ym}\,y_n)] \right|^2 - M \right\}, \qquad M < N \tag{7.30}$$

β being a Lagrange multiplier. By maximizing Eq. (7.30) relative to all the b_n (real) Wernecke and D'Addario found it convenient to select a value of β for the maximization procedure. This determined the relative importance of h_τ^{ME} and the noise residual given by the $\sum_m$ term. Such a treatment of noise follows the prescription of Eq. (1.33b) in Sec. 1.4. They initally chose β assuming one datum at $k_{x1} = k_{y1} = 0$ and a uniformly gray ME object. For this the ME procedure yielded

$$\beta_1 = \frac{A\sigma_1^2}{2(b_1 + \delta_1)\delta_1} \qquad \text{for } \delta_1 = |\,Ab_1 - S_1\,| = \sigma_1 \tag{7.31}$$

Wernecke and D'Addario found for their problems of interest that this value for β was adequate to maintain inequality Eq. (7.29) in the process of iterating to a solution of all the equations for the b_n obtained by setting $\delta F = 0$ in Eq. (7.30). They observed there could be no value of β which would insure

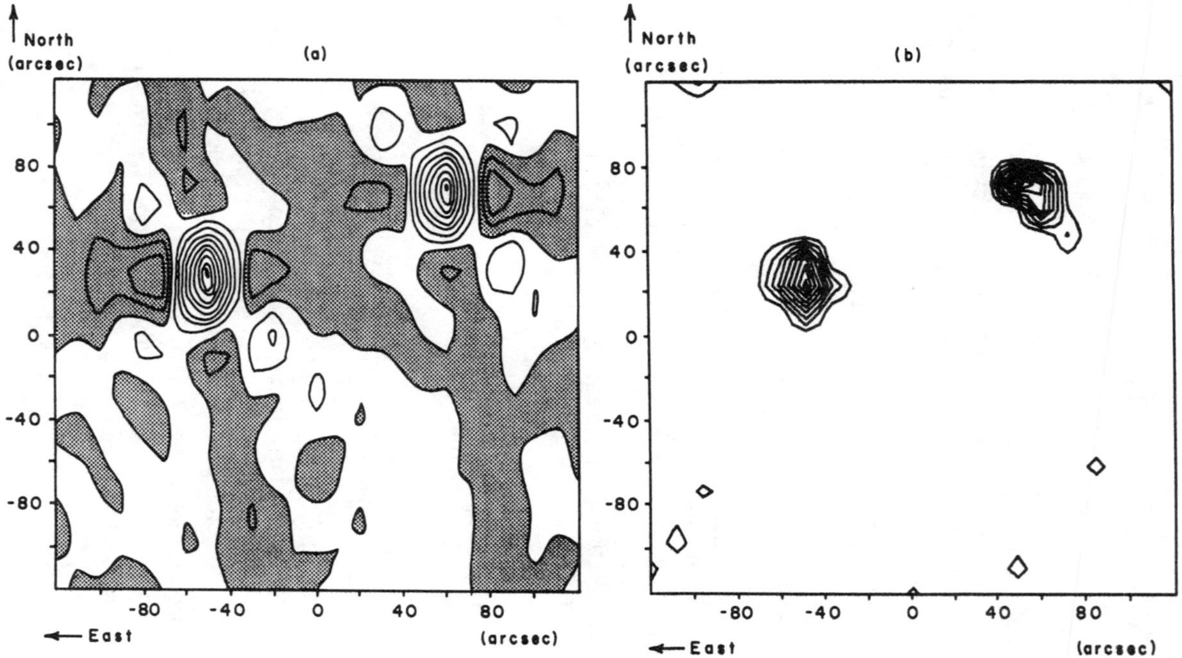

Figure 7.5 Reconstruction of Cygnus A given by (a) direct Fourier transformation of all measurements by Eq. (7.30). The shaded areas have negative brightness. (b) after 23 iterations of the ME algorithm based on Eq. (7.28). Error Eq. (7.31) was 0.157. All contours are drawn at intervals of 10 percent of maximum brightness. From [7.5] Fig. 11, © 1977, IEEE.

satisfaction of Eq. (7.29) if one or more measurements were unduly noisy or if some of the σ_m variances were assumed unrealistically small by an optimistic observer.

A dramatic example of the improvement attained by this ME reconstruction over conventional Fourier inversion is illustrated by the radio double star Cygnus A (3C405), which has $\langle n \rangle / z \cong 10^6$ at 960 megahertz. The Stanford University interferometer [7.7] operating at a wavelength of nearly 2.8 cm (10.7 gigahertz) took essentially continuous measurements over nine concentric ellipses in the k_x-k_y plane. All these data were employed in a direct Fourier transform program to obtain the $\{b_n\}$ but only discrete data points at 30° intervals on the innermost five k-ellipses were furnished to the ME algorithm.

The Fourier program reconstructed the relative-intensity brightness object $b(x, y)$ as a continuous function of x and y from a discrete version of the mate of Eq. (7.28),

$$b(x, y) = \sum_{m}^{M} \alpha_m \text{ Real } \left\{ S_m \text{ (measured) } \exp[j(k_{xm}x + k_{ym}y)] \right\} \quad (7.32)$$

The α_m are constants which can be varied to improve subjectively the appearance of the radiating object, analogous to the use of "lag windows" in orthodox power spectrum estimation.

Figure 7.5 shows vividly the improvement in the reconstructed Cygnus A object by this ME method after 23 iterations. The background nonphysical brightness area of *negative* sign, shown shaded in the Fourier reconstruction, often results from using Eq. (7.32) with incomplete wavenumber $\{k_x, k_y\}$ data. Such a nonphysical area cannot occur in an ME reconstruction.

The relative rms error, defined as

$$\text{error} = \left[\frac{\sum_m | S_m \text{ (measured) } - S_m^{ME}(k_{xm}, k_{ym}) |^2}{\sum_m | S_m \text{ (measured) } |^2} \right]^{1/2} \quad (7.33)$$

took the value 0.157.

REFERENCES

[7.1] B.H. Soffer and H. Kikuchi, Maximum entropy image estimation. Analysis of image formation with thinned random arrays, Annual Report, 15 Feb. 1978–31 Aug. 1980, Hughes Research Laboratories, Malibu, CA, 90265, Feb. 1981, Section 4.B.

[7.2] R. Kikuchi and B.H. Soffer, Maximum entropy image restoration I. The entropy expression, *Jour. Opt. Soc. Am.*, *67*, Dec. 1977, pp. 1656–1665.

[7.3] J. Klauder and E.C.G. Sudarshan, *Fundamentals of Quantum Optics*, New York: Benjamin, 1968, p. 138.

[7.4] J.P. Burg, Maximum entropy spectral analysis, In *Modern Spectrum Analysis*, D.G. Childers (ed.). IEEE Press Book 1978, pp. 34–41.

[7.5] S.J. Wernecke and L.R. D'Addario, Maximum entropy image reconstruction, IEEE Trans. Computers, *C-26*, April 1977, pp. 351–364. Reprinted in *Modern Spectrum Analysis*, D.G. Childers (ed.). IEEE Press Book, 1978, pp. 287–300.

[7.6] J. Capon, High resolution frequency-wavenumber spectrum analysis, Proc. IEEE, *57*, Aug. 1969, pp. 1408–1418. Reprinted in *Modern Spectrum Analysis*, D.G. Childers (ed.). IEEE Press Book, 1978, pp. 119–129.

[7.7] R.N. Bracewell et al., The Stanford five-element radio telescope, Proc. IEEE, *61*, Sept. 1973, pp. 1249–1257.

8

Reliability-Risk Maximum Entropy Analysis

8.1 INTRODUCTION

This chapter describes a system in a general sense as a network of branches (paths) interconnected at nodes (components). Various paths through the network must function or the system fails. Each path is defined by alternating nodes and branches, all connected. Paths may share components and we *assume* that the components fail independently and the failure of any one component causes failure of any path passing through it.

In Sec. 8.2 we treat the problem of estimating the failure probability of a single component under increasing stress from incomplete data. The cumulative failure probability $P_F(x)$ is defined as the fraction of an ensemble of that component which would fail as the stress is increased from zero to x. We deduce an ME curve of $P_F(x)$ versus x from tests, in which a large number of that component has been divided into various groups and each group is stressed to a different level, x. We presume the following data: a curve of the cumulative fraction $P_s(x)$ stressed from zero to x and the net probability of failure, P_{fail}, for all the groups stressed to different levels. From these data we deduce a Boltzmann curve of $P_F(x)$ versus x.

The probability of system failure is related to the probabilities of various path failures in Sec. 8.3. Each path failure probability depends on its component failure probabilities. Generally, the problem of assessing system failure probability from path failure probabilities is time-consuming because the paths are usually defined to share components in a convenient way and are not independent of each other. However, if a logical procedure is followed to define a complete set of independent paths the computation of system failure probability simplifies to a sum of the failure probabilities of *these* paths.

In Sec. 8.4 we begin discussing the estimation of various component failure probabilities from path failure probability data. The Boltzmann expression for ME nonfailure probability of a component is of Fermi-Dirac form, convenient for computation. But, the ME values of these component nonfailure probabilities are often not close to their assumed values in synthetic-data examples without a priori bounds on the values.

Therefore, in Sec. 8.5 the corresponding Gibbs expression for the expected values $\{\langle q_k \rangle\}$ is developed for the nonfailure probabilities of the components, with bounds on each q_k. These, too, are more useful for tighter bounds.

In Sec. 8.6 this ME-Gibbs analysis is applied to compute the expected values $\langle -\log(q_k)\rangle$ and $\langle q_k \rangle$ from path nonfailure data when we know that all $\{q_k\}$ are close to 1. A synthetic data example suggests that this method could be useful for selecting a limited set of paths in a system which would give ME estimates closest to the actual component nonfailure probability values, on the basis of computations for a model of the system. A relatively large deviation of an ME nonfailure probability of a system component and the desired value would indicate the need for further tests of that component.

8.2 PROBABILITY OF FAILURE FOR A COMPONENT UNDER STRESS

A problem of great importance to manufacturers and buyers of a product line is the estimation of how the failure probability of the product increases with increasing stress of some sort. To illustrate the general features of failure-stress analysis we will consider specifically an ensemble of light bulbs which differ among themselves by variations in the manufacturing process. We choose to measure stress on the bulbs by the time in use. We assume the manufacturer has designed the bulbs to have a maximum working life of x_M hours; that is, they will all fail in this time.

The question we address is: What is the cumulative probability of failure, $P_F(x)$—also called *fragility*—of a new bulb operated for a specified number of hours, x? $P_F(x)$ is defined as the fraction of new bulbs which have failed at various times during x hours of operation. We further assume, for simplicity, that $P_F(x)$ depends only on the total operating time x independently of the number and duration of the "on" and "off" intervals during time x.

Figure 8.1 illustrates qualitatively this cumulative failure probability, $P_F(x)$ versus stress time x up to the maximum working time x_M when they are all presumed to have failed. We shall predict the shape of this curve

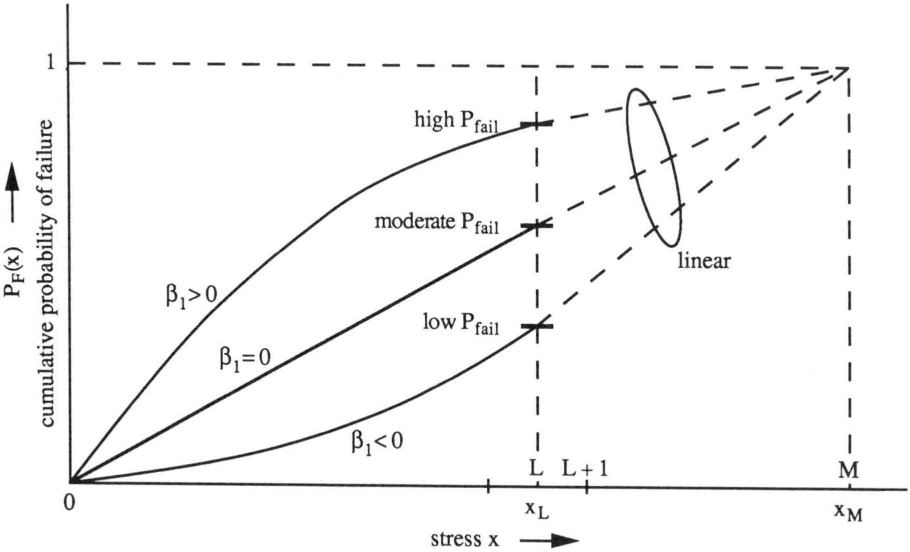

Figure 8.1 Cumulative probability of failure, $P_F(x)$, of an ensemble of objects as a function of stress, x. $P_F(x) = 1$ (all objects have failed) at maximum stress, x_M. The shape of the ME curve as indicated by its Lagrange multiplier value is derived from a probability-of-failure datum P_{fail} reported by the manufacturer from a series of tests based on a curve of cumulative probability of stress, $P_s(x)$ like that shown in Fig. 8.2.

from a single failure-probability datum, P_{fail}, obtained from a test curve of the relative number of bulbs stressed versus stress level (Fig. 8.2).

Assume in Fig. 8.1 a representative number of new bulbs is stressed from $x = 0$ to $x_1 - dx/2$. The fraction which has failed along the way is $P_F(x_1 - dx/2)$. If the stress on the unfailed number is increased further to $x_1 + dx/2$ the total failed fraction will increase to $P_F(x + dx/2) \cong P_F(x_1 - dx/2) + [dP_F(x)/dx]_{x_1}\, dx$. Here, $dP_F(x)/dx = p_F(x)$ is the rate of change of failure probability with increasing stress x.

Ideally, the manufacturer will have failure data for one large, statistically significant, group of bulbs stressed from 0 to x_M, or else data for many smaller representative groups stressed to various levels between 0 and x_M. We shall assume, for proprietary reasons, that he divulges only one failure datum, P_{fail}, based on a curve, $P_s(x)$ versus x, of a cumulative fraction of test bulbs stressed up to $x_L < x_M$. This test curve appears in Fig. 8.2. A typical dot on the curve at x indicates that he stressed a group of bulbs consisting of a fraction $P_s(x)$ of all test bulbs from stress 0 to x. Of this fraction, we would expect a fraction $P_F(x)$ to have failed—if we knew P_F. The manufacturer doesn't tell us how many of *each* group of test bulbs failed

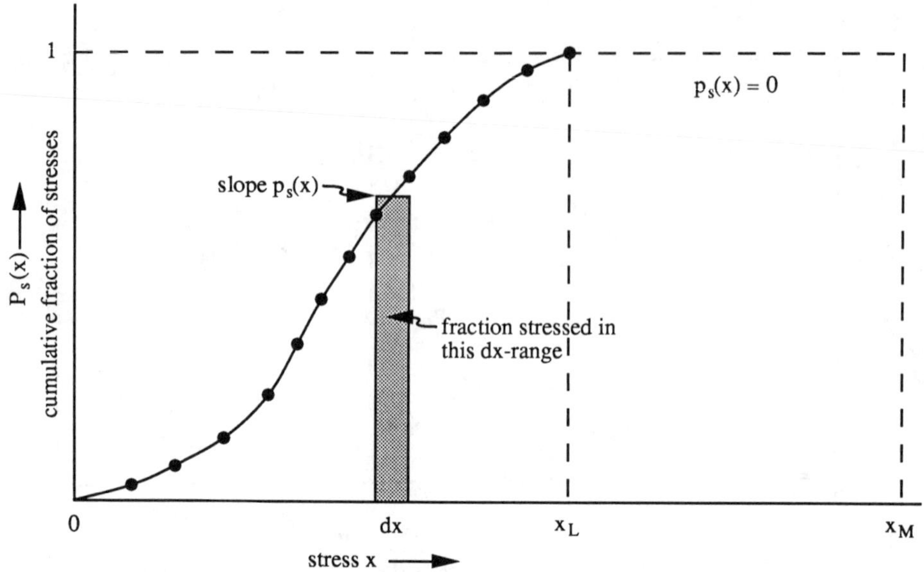

Figure 8.2 Cumulative probability of stress, $P_s(x)$, reported by the manufacturer for different groups of bulbs tested independently of the other groups. Each dot on the curve represents a group of bulbs tested from stress zero to the value x indicated by the dot. No bulbs are stressed beyond x_L. P_{fail} is the total probability of failure reported for all the groups tested.

in its range $(0, x_L)$, only the total failure probability, P_{fail}, for *all* the groups. From this one datum, we intend to construct an ME curve for $P_F(x)$ from $x = 0$ all the way to x_M.

To relate P_{fail} to the P_F and P_s curves, we proceed as follows. On Fig. 8.2 a fraction $dP_s(x) = (dP_s/dx)_x\, dx = p_s(x)\, dx$ of the test bulbs have been stressed from zero to this narrow range $(x \pm dx/2)$. Of these, an (unknown) fraction $P_F(x)$ have failed somewhere in the range $(0, x)$. The fraction of all the test bulbs which have failed in the dx range is approximately $P_F(x)p_s(x)\, dx$ if dx is small. The total fraction of all the test bulbs which fail over the entire test range $(0, x_L)$ is the integral of $P_F(x)p_s(x)\, dx$. This is the P_{fail} datum,

$$P_{fail} \text{ datum} = \int_0^{x_L} P_F(x)p_s(x)\, dx = \int_0^{x_M} P_F(x)p_s(x)\, dx \tag{8.1}$$

We extend the range of integration to x_M for convenience, $p_s(x)$ being zero over (x_L, x_M).

If the tester had merely increased the stress from zero to x_L on one large group of bulbs and recorded the failure probability $P_F(x_L)$, we would have a weaker constraint than Eq. (8.1).

We now integrate Eq. (8.1) by parts to obtain a constraint on $p_F(x)$,

$$P_{\text{fail}} = P_F(x)P_s(x)\Big|_0^{x_M} - \int_0^{x_M} P_s(x)p_F(x)\,dx = 1 - \int_0^{x_M} P_s(x)p_F(x)\,dx \tag{8.2}$$

It is convenient to formulate the Boltzmann entropy for discrete levels $p_F(x_i) = p_{Fi}$, so we discretize Eq. (8.2) as

$$1 - \sum_{i=1}^{M} P_{si}p_{Fi} - P_{\text{fail}} = 0 \tag{8.3}$$

The second constraint on p_F is the obvious one,

$$\int_0^{x_M} p_F(x)\,dx = P_F(x_M) = 1 \tag{8.4}$$

which discretizes to

$$\sum_{i=1}^{M} p_{Fi} - 1 = 0 \tag{8.5}$$

To compute the ME set of numbers $P_{F1}, \ldots, P_{FM}$ in M "bins" we imagine a large number N of bulbs. As we increase the stress on them first $n_1 = p_{F1}$ N of them would fail and be put into bin 1, then $n_2 = p_{F2}$ N more of them would fail and be put into bin 2, and so on, until $n_M = p_{FM}$ N fail in the last bin M. $\sum n_i = N$ because all have failed by assumption when stress x reaches x_M. The number of distinct ways of realizing a set of numbers $\{n_i\}$ is just the number of combinations C of Eq. (1.3),

$$C = \frac{N!}{\prod_n n_i!}, \qquad \sum n_i = N \tag{8.6}$$

Our objective function is just log(C) plus constraints Eqs. (8.3) and (8.4), each multiplied by a Lagrange multiplier,

$$F = \log C + \beta_1'\Big(1 - \sum_{i=1}^{M} P_{si}p_{Fi} - P_{\text{fail}}\Big) + \beta_2'\Big(\sum_i p_{Fi} - 1\Big) \tag{8.7}$$

From Eq. (1.8) we have

$$\log C = -\sum_i^{M} n_i \log\left(\frac{n_i}{N}\right) = -N\sum_i^{M} p_{Fi}\log(p_{Fi}) \tag{8.8}$$

Upon setting a variation $\delta F = 0$ for each δp_{Fi} we obtain, with the substitutions $\beta_1 = \beta_1'N$ and $\beta_2 = \beta_2'N + 1$,

$$p_{Fi}^{ME} = \exp(-\beta_1 P_{si} - \beta_2), \qquad 1 \le i \le M \tag{8.9}$$

In the limit of continuous x we take the limit of Eq. (8.9) to be

$$p_F(x)^{ME} = \exp[-\beta_1 P_s(x) - \beta_2], \qquad 0 \le x \le x_M \tag{8.10}$$

Then the ME cumulative probability of failure, $P_F^{ME}(x)$, in Fig. 8.1 is

$$P_F(x)^{ME} = \int_0^x p_F(y)^{ME}\, dy = e^{-\beta_2} \int_0^x e^{-\beta_1 P_s(y)}\, dy \tag{8.11}$$

β_2 may be expressed in terms of β_1 by constraint Eq. (8.4),

$$\int_0^{x_M} p_F(x)\, dx = e^{-\beta_2} \int_0^{x_M} e^{-\beta_1 P_s(y)}\, dy = 1 \tag{8.12}$$

Equation (8.2) enables us to solve implicitly for β_1 from

$$P_{fail} = 1 - \frac{\left[\int_0^{x_M} P_s(x) e^{-\beta_1 P_s(x)}\, dx\right]}{\left[\int_0^{x_M} e^{-\beta_1 P_s(x)}\, dx\right]} \tag{8.13}$$

For small P_{fail} $\beta_1 < 0$ and $P_F(x)$ is relatively small over the entire range of x in Fig. 8.1. For large P_{fail}, $\beta_1 > 0$ and the entire P_F curve lies above the linear curve for $\beta_1 = 0$.

The ME solution predicts quite reasonably that with no test data in the range (x_L, x_M) the rate of failure $p_F(x)$ in this range is constant with increasing stress and so $P_F(x)$ rises linearly in Fig. 8.1.

8.3 PROBABILITY OF A SYSTEM FAILURE RELATIVE TO ITS PATH FAILURE PROBABILITIES

A complex system should sometimes be tested for reliability by computing or measuring the probabilities of failure along various paths through the system. These probabilities depend on the failure probabilities of the path components. A good example of such a system is a computer, which routes data through a network of nodes (components) in the most expedient way. Maximum entropy estimation of component failure probabilities from relatively few path failure probabilities can yield "most probable" failure probabilities of more of the crucial paths, and therefore, estimate the overall system probability of failure.

A portion of the network diagram for a large system is shown in Fig. 8.3. Here, various paths such as T_i traverse the system, passing through components such as C_l. The paths are not all independent because many

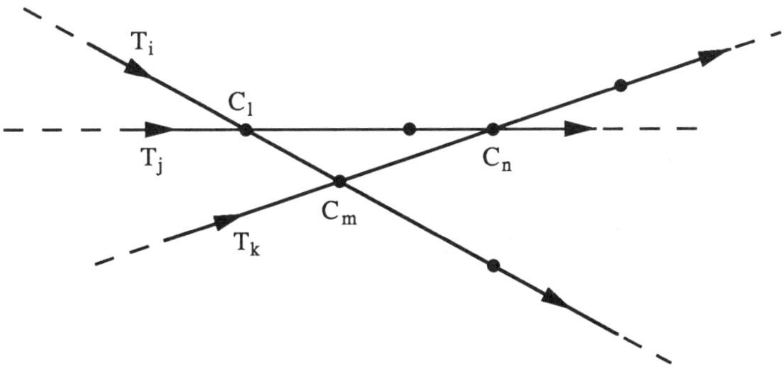

Figure 8.3 A portion of the network diagram for a large system. C_l, etc., are components; T_i, etc., represent paths through the network. Each dot indicates a component on a path.

share components. System failure probability is a function of all such path failure probabilities.

The relationships between path and component failures can also be summarized by the matrix of Fig. 8.4. There the paths are represented by rows; the components, by columns. m_{ij} is an entry indicating the significance of component C_j in path i. T_i passes through C_j if $m_{ij} = 1$; if $m_{ij} = 0$ path i is independent of C_j. The probability of failure of C_j is denoted p_j and is defined for an ensemble of systems. That is, p_j is the fraction of a large number of similar systems in which C_j fails during a period of time the system is expected to be reliable. We shall assume that a component has either failed or not failed during operation and, if it has failed, all paths through it

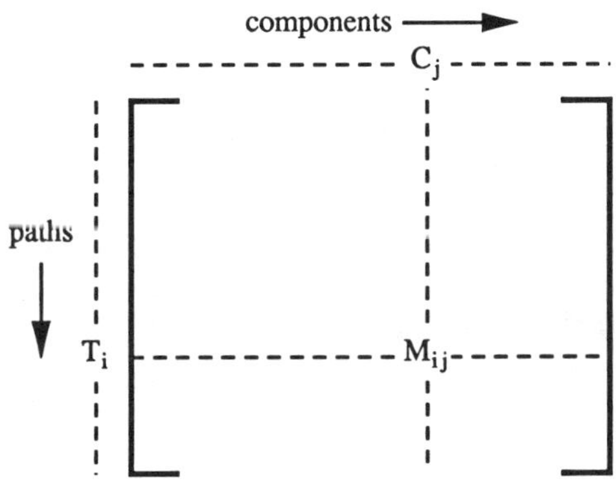

Figure 8.4 Connection matrix for the paths in a system network, part of which is shown in Fig. 8.3. $m_{ij} = 1$ only if path i passes through component C_j and is zero otherwise.

have failed. We further assume that the components fail independently of each other, which may not be true if the same stress(es) act on many of them.

Path failure data will be incomplete for assessing system probability of failure if component failure probabilities are not all precisely known. This can occur, for example, when the components are too numerous to have their failure probabilities computed accurately and economically from the required number of path failure data. We will present an ME methodology for estimating the unknown component path failure probabilities from relatively few measured path failure probabilities in Fig. 8.3. Then the known and ME-estimated component failure probabilities will enable one to assess more accurately all the path failure probabilities, and thereby, determine the system reliability.

Incidentally, the problem of calculating the probability of large-system failure from the failure probabilities of paths as they are usually conveniently defined is not a simple one, for two major reasons. First, paths may not be mutually "disjoint"; every path may share one or more components (nodes) with other paths, and this makes the path computation problem NP-hard (solution time increases exponentially with the number of components). Since system probabilities can be computed only from path probabilities this computation, too, is NP-hard. Second, even when paths are disjoint the components on various paths may be statistically dependent, and the computation of path probabilities may require evaluation of high-dimensional joint probability integrals, another NP-hard problem.

Fortunately, some approximation methods have been developed which reduce the computational complexity. Whereas conventional methods of computing path failure probabilities require NP-hard computational times ([8.1] has a good survey) new methods compute a complete set of independent paths through the system and require computational times which increase only as the cube of the number of components [8.2]. System probability of failure is a simple sum of the failure probabilities of these paths. We shall not pursue this matter further but rather concentrate on the ME-estimation of component failure probabilities from limited path failure probability data.

8.4 ME-BOLTZMANN ESTIMATION OF COMPONENT FAILURE PROBABILITIES FROM PATH FAILURE DATA

Path T_i in Fig. 8.3 can fail if any component on it fails, but there is only one way it will *not* fail: if all its components do not fail. In terms of this path failure probability P_i and its nonfailure probability $Q_i = 1 - P_i$ and all the component $\{p_j\}$ and $\{q_j\}$, $q_j = 1 - p_j$, the nonfailure probability Q_i is

$$Q_i = \left(\sum_{k=1}^{K} m_{ik} q_k\right)\left(\prod_l{}' m_{il} q_l\right), \qquad i \le I, \text{ number of paths} \tag{8.14}$$

Here, $m_{ik} = 1$ if C_k lies on path i and is zero otherwise. Equation (8.14) furnishes $I < K$ constraints on the K unknown q_k via all the measured Q_i. The q_l in the primed product represents a known nonfailure probability, if there are any, of component C_l.

The Boltzmann entropy of all the unknown components is the *sum* of the individual entropies of the components because they are assumed to fail independently. The entropy of component C_k is derived in Appendix E as

$$h_k = -(p_k \log p_k + q_k \log q_k), \qquad q_k = 1 - p_k \tag{8.15}$$

This follows from the number of combinations of N components with N_k in a failure state and $(N - N_k)$ in a nonfailure state.

The objective function F is written in terms of the q_k as

$$F = -\sum_k [(1 - q_k)\log(1 - q_k) + q_k \log q_k] + \sum_{i=1}^{I} \beta_i \left[\prod_{k=1}^{K} m_{ik} q_k - Q_i'\right] \tag{8.16}$$

where the known $Q_i' = Q_i/(\prod_l m_{il} q_l)$ from Eq. (8.14) and β_i is the ith Lagrange multiplier. When we set a variation $\delta F = 0$ for arbitrary δq_k in Eq. (8.16) we obtain straightforwardly

$$q_k^{\text{ME}} = \frac{\exp\left[(q_k^{\text{ME}})^{-1} \sum_i^{I} \beta_i m_{ik} Q_i'\right]}{1 + \exp\left[(q_k^{\text{ME}})^{-1} \sum_i^{I} \beta_i m_{ik} Q_i'\right]} < 1, \qquad k \le K \tag{8.17}$$

This is the Fermi-Dirac form Eq. (1.21), and we see $0 < q_k^{\text{ME}} < 1$. Notice if C_k lies on none of the I measured paths, all $m_{ik} = 0$, the numerator is 1 and $q_k^{\text{ME}} = 1/2 = p_k^{\text{ME}}$. The K equations Eq. (8.17) must be solved self-consistently for the K unknown q_k^{ME} and I unknown β_i, so as to satisfy the I constraints Eq. (8.14). In the iterative procedure analogous to that of Sec. 3.3 it is convenient to update the β_i as $\beta_i + \Delta\beta_i$ each iteration by solving the set of equations

$$Q_i' - \hat{Q}_i' = \sum_{j=1}^{I} \frac{\partial \hat{Q}_i'}{\partial \beta_j} \Delta\beta_j, \qquad i \le I \tag{8.18}$$

the caret denoting a current estimate, and

$$\frac{\partial \hat{Q}_i'}{\partial \beta_j} = \hat{Q}_i' \left[\sum_k m_{ik} m_{jk} \frac{\hat{p}_k/\hat{q}_k}{1 + \hat{p}_k \log(\hat{q}_k/\hat{p}_k)} \right] \hat{Q}_j' = \frac{\partial \hat{Q}_j'}{\partial \beta_i} \tag{8.19}$$

The sum includes component C_k only if it lies on both paths i and j.

We have written a numerically well-behaved computer code to evaluate the nonfailure probabilities q_k^{ME}. However, for some synthetic-data connection matrices of Fig. 8.4, the $\{q_k^{\mathrm{ME}}\}$ did not correspond closely to the assumed $\{q_k\}$, which determine the synthetic data $\{Q_i\}$ from which the $\{q_k^{\mathrm{ME}}\}$ are computed. In a typical synthetic-data problem some of the $\{q_k^{\mathrm{ME}}\}$ were considerably lower than assumed. This is plausible considering that entropy h_k of Eq. (8.15) for component C_k tends toward its maximum at $p_k = q_k = 1/2$.

The tendency of this Boltzmann method to predict inaccurately nonfailure probabilities in synthetic-data examples does not constitute a fundamental objection to the ME method. It merely means that the $\{q_k\}$ assumed in each of these examples are "far away" from the most probable $\{q_k^{\mathrm{ME}}\}$ which yield the synthetic data $\{Q_i\}$. For more accurate ME estimation we must restrict the range of possible values for the $\{q_k\}$ in a realistic way.

The ME-Boltzmann method suffers the disadvantage of not allowing convenient restriction of the $\{q_k\}$ to specified ranges. Therefore, we turn to the ME-Gibbs method, which is suited to this task.

8.5 ME-GIBBS ESTIMATION OF COMPONENT FAILURE PROBABILITIES FROM PATH FAILURE PROBABILITIES

The Gibbs method is based on an underlying probability distribution for the values of the unknowns being estimated, in this case the $\{q_k\}$. To avoid confusion in discussing a probability distribution for a probability we redefine the q_k as $b_k = q_k$. The probability distribution for b_k is denoted $\mathcal{P}_k(b_k)$—not to be confused with failure probability P_k for path k.

The Boltzmann entropy h_k of Eq. (8.15) is converted to Gibbs form as follows. The "point" nonfailure probability $q_k = b_k$ is replaced by $\mathcal{P}(b_k)\,\Delta b_k$, the relative number of C_k components within the range $(b_k \pm \Delta b_k/2)$, and p_k is replaced by $1 - \mathcal{P}_k(b_k)\,\Delta b_k$. Then, Eq. (8.15) is summed over b_k from $b_{k1} > 0$ to $b_{k2} \leq 1$. The Gibbs entropy h_{Gk} for C_k is, therefore,

$$\begin{aligned} h_{Gk} = -\sum_{b_{k1}}^{b_{k2}} \Big\{ &\mathcal{P}_k(b_k)\,\Delta b_k \log[\mathcal{P}_k(b_k)\,\Delta b_k] \\ &+ [1 - \mathcal{P}_k(b_k)\,\Delta b_k] \log[1 - \mathcal{P}_k(b_k)\,\Delta b_k] \Big\} \end{aligned} \tag{8.20}$$

Δb_k is a small interval we will let approach zero in a suitable manner. Since

$$\log[1 - \mathcal{P}(b_k)\,\Delta b_k] \approx \mathcal{P}(b_k)\,\Delta b_k + \mathcal{O}(\Delta b_k)^2 \tag{8.21}$$

we neglect terms of order $(\Delta b_k)^2$ in Eq. (8.20) and obtain for it

$$h_{Gk} = -\sum_{b_{k1}}^{b_{k2}} \mathcal{P}_k(b_k)\,\Delta b_k \log[\mathcal{P}_k(b_k)] + \sum \mathcal{P}_k(b_k)\,\Delta b_k[1 - \log(\Delta b_k)] \tag{8.22}$$

Now if we let Δb_k be uniform over the range (b_{k1}, b_{k2}) we can extract the factor $[1 - \log(\Delta b_k)]$ from the second sum. When $\Delta b_k \to 0$ in the remaining terms we obtain for the total Gibbs entropy $h_G = \sum_k h_{Gk}$ from Eq. (8.22)

$$h_G = -\sum_k \int_{b_{k1}}^{b_{k2}} \mathcal{P}_k(b_k) \log[\mathcal{P}_k(b_k)]\, db_k + [1 - \log(\Delta b_k)] \sum_k \int_{b_{k1}}^{b_{k2}} \mathcal{P}_k(b_k)\, db_k \tag{8.23}$$

We will show that the factor $[1 - \log(\Delta b_k)]$ does not affect $\mathcal{P}_k(b_k)$ because it can be absorbed into a Lagrange multiplier.

Each constraint Eq. (8.14) is now written to replace the unknown q_k by its expected value $\langle q_k \rangle = \langle b_k \rangle$ over its probability $\mathcal{P}_k(b_k)$,

$$\mathrm{Q}_i = \prod_k^{\mathrm{K}} m_{ik}\left[\int_{b_{k1}}^{b_{k2}} b_k \mathcal{P}_k(b_k)\, db_k\right]\Big(\prod_l{}' m_{il} b_l\Big), \qquad i \le \mathrm{I}, \tag{8.24}$$

subject to the normalization constraint

$$\int_{b_{k1}}^{b_{k2}} \mathcal{P}_k(b_k)\, db_k = 1 \tag{8.25}$$

The Gibbs objective function $\mathrm{F_G}$ replacing Eq. (8.16) is

$$\mathrm{F_G} = h_G - \sum_{i=1}^{\mathrm{I}} \beta_i\left[\prod_k^{\mathrm{K}} m_{ik} \int_{b_{k1}}^{b_{k2}} b_k \mathcal{P}_k(b_k)\, db_k - \mathrm{Q}_i'\right] + \sum_k \gamma_k\left[\int_{b_{k1}}^{b_{k2}} \mathcal{P}_k(b_k)\, db_k - 1\right] \tag{8.26}$$

For an arbitrary variation $\delta \mathrm{F_G} = 0$ relative to any of the $\delta\mathcal{P}_k(b_k)$ at any value of b_k we obtain

$$0 = -\log[\mathcal{P}_k(b_k)] - 1 + [1 - \log(\Delta b_k)] - \sum_i^{\mathrm{I}} \beta_i (m_{ik} b_k)\Big(\prod_{k' \neq k} m_{ik'} \langle b_{k'} \rangle\Big) + \gamma_k \tag{8.27}$$

Now the term, $-\log(\Delta b_k) + \gamma_k$ can be redefined γ_k' and $\exp(\gamma_k')$ denoted A_k, with the result

$$\mathcal{P}_k(b_k) = A_k \exp(-L_k b_k), \qquad L_k = \frac{1}{\langle b_k \rangle} \sum_i \beta_i m_{ik} Q_i' \tag{8.28}$$

A_k is evaluated to satisfy the normalization constraint within the last [ ] of Eq. (8.26) as

$$A_k = L_k(e^{-x_1} - e^{-x_2})^{-1}, \qquad x_1 = L_k b_{k1},\, x_2 = L_k b_{k2} \tag{8.29}$$

The expected value $\langle b_k \rangle$ of b_k, the integral in Eq. (8.24), evaluates to

$$\langle b_k \rangle = L_k^{-1}(e^{-x_1} - e^{-x_2})^{-1}[(x_1 + 1)e^{-x_1} - (x_2 + 1)e^{-x_2}] \tag{8.30}$$

[Compare with Eq. (8.17).]

The reduced Gibbs entropy h_G', consisting of only the first sum in Eq. (8.23), which determines probability distribution $\mathcal{P}(b_k) = \Pi\mathcal{P}_k(b_k)$, evaluates to

$$h_G' = -\sum_k \log A_k + \sum_k L_k \langle b_k \rangle \tag{8.31}$$

The self-consistent solution for all the A_k and L_k requires iteration because all the β_i must yield all the L_k of Eq. (8.28) needed to evaluate the $\langle b_k \rangle$ of Eq. (8.30) so as to satisfy all the I constraints of Eq. (8.24).

An algorithm very similar to the geotomographic one of Sec. 3.3 was written to converge to a self-consistent solution starting from arbitrary starting values $\{\boldsymbol{\beta}^{(0)}\}$. Convergence was quite rapid and independent of those starting values. However, the $\{b_k^{\mathrm{ME}}\}$ were not always close to the assumed values in synthetic-data problems when the b_k were unrestricted over the range (0, 1). This behavior is similar to that of the Boltzmann solutions described in Sec. 8.4.

We conclude that generally the Gibbs ME estimation of component nonfailure probabilities will only be useful if their ranges are restricted by prior knowledge to be rather narrow. For many problems of interest the systems are required to be highly reliable, so the component and path nonfailure probabilities are all nearly 1. We treat this situation in the next section.

8.6 ME-GIBBS ESTIMATION OF COMPONENT NONFAILURE PROBABILITIES CLOSE TO 1

In Sec. 8.5 we redefined q_k as b_k, the nonfailure probability of component C_k, and derived the Gibbs probability distribution $\mathcal{P}_k(b_k)$. Now we want to define b_k as $-\log(q_k) = -\log(1 - p_k) = p_k + p_k^2/2 + \ldots \simeq p_k + p_k^2/2$ for $q \gtrsim 0.98$. This definition avoids some complications in the ME algorithm and makes the constraints assume a simpler form, analogous to Eq. (3.3) for geotomography,

$$\begin{aligned} Y_i &= -\log Q_i' = \sum_k^K m_{ik}\langle -\log q_k\rangle \\ &\approx \sum_k m_{ik}\langle p_k + \frac{1}{2} p_k^2\rangle = \sum_k m_{ik}\langle b_k\rangle \end{aligned} \tag{8.32}$$

The Gibbs entropy h_G of Eq. (8.23), derived in Sec. 8.5 for $b_k = q_k$, can be easily rederived for $b_k = -\log(q_k) \simeq p_k + \frac{1}{2}p_k^2$ because the starting equation was Eq. (8.15) for the Boltzmann entropy, symmetrical in p_k and q_k. Now we replace p_k by $b_k - p_k^2/2 \simeq b_k - b_k^2/2$ in Eq. (8.15), then replace this by $\mathcal{P}(b_k)\,\Delta b_k - [\mathcal{P}(b_k)\,\Delta b_k]^2/2$ for the Gibbs h_{Gk}, neglect terms of order $(\Delta b_k)^2$, and arrive at Eq. (8.20) again. Therefore, h_G of Eq. (8.23) is the proper Gibbs entropy, in terms of $b_k = -\log(q_k)$. Constraint Eq. (8.25) remains the same.

The F-function Eq. (8.26) remains the same except that the first bracketed expression is replaced by

$$\left[\sum_k^K m_{ik} \int_{b_{k1}}^{b_{k2}} b_k \mathcal{P}_k(b_k)\, db_k - Y_i\right] \tag{8.33}$$

from Eq. (8.32). By the δF-variational procedure we obtain, in place of Eq. (8.28),

$$\mathcal{P}_k(b_k) = A_k e^{-L_k b_k}, \qquad L_k = \sum_i^I \beta_i m_{ik} \tag{8.34}$$

This is the Gibbs counterpart of the Boltzmann Eq. (3.10).

In terms of the new L_k, A_k of Eq. (8.29) again normalizes $\mathcal{P}_k(b_k)$, where, now, $b_{k1} = -\log(q_{k2})$, $b_{k2} = -\log(q_{k1})$ and $q_{k1} < q_{k2}$. $\langle b_k\rangle$ is given by Eq. (8.30), in terms of the new $x_1 = L_k b_{k1}$ and $x_2 = L_k b_{k2}$.

A Gibbs algorithm was written, very similar to the one described in Sec. 3.3 for a Boltzmann geotomographic distribution. Starting from an initial choice of $\beta_1^{(0)} = \ldots = \beta_I^{(0)}$ the L_k are computed by Eq. (8.34), then the $\langle b_k\rangle$ by Eq. (8.30), and the $\hat{Y}_i$ estimates by the right side of Eq. (8.32). If these differ too much from the true Y_i, the β_j are all updated by solving the matrix equations

$$Y_i - \hat{Y}_i = \sum_{j=1}^{I} \frac{\partial \hat{Y}_i}{\partial \beta_j} \Delta\beta_j, \qquad i \le I \tag{8.35}$$

where

$$\frac{\partial \hat{Y}_i}{\partial \beta_j} = \sum_k m_{ik} m_{jk} \left[-L_k^{-2} + \left(\frac{b_{k1}e^{-x_1} - b_{k2}e^{-x_2}}{e^{-x_1} - e^{-x_2}} \right)^2 - \left(\frac{b_{k1}^2 e^{-x_1} - b_{k2}^2 e^{-x_2}}{e^{-x_1} - e^{-x_2}} \right) \right] = \frac{\partial \hat{Y}_j}{\partial \beta_i} \tag{8.36}$$

Then, $\beta_j \rightarrow \beta_j + \Delta\beta_j$ for each $j \leqq I$ and the algorithm recomputes the L_k and repeats the loop of computations. When **Y** is sufficiently close to $\hat{\mathbf{Y}}$ the $\langle q_k \rangle^{ME}$ are computed by Eqs. (8.40) and (8.41) below.

$\langle q_k \rangle^{ME}$ is obtained from $\langle b_k \rangle^{ME} = \langle -\log q_k \rangle^{ME}$ by expanding the log-term for $q_k \gtrsim 0.98$ as

$$b_k = -\log q_k \approx p_k + \frac{1}{2} p_k^2$$
$$\text{so } \langle b_k \rangle \approx \langle p_k \rangle + \frac{1}{2} \langle p_k^2 \rangle \tag{8.37}$$

We obtain sufficient accuracy by neglecting terms of order p_k^3 and higher. b_k^2 is approximately

$$b_k^2 \approx p_k^2 + p_k^3 \approx p_k^2$$
$$\text{so } \langle b_k^2 \rangle \approx \langle p_k^2 \rangle \tag{8.38}$$

From the expectations in Eqs. (8.37) and (8.38) we obtain for $\langle q_k \rangle^{ME}$

$$\langle q_k \rangle^{ME} \approx 1 - \langle p_k \rangle^{ME} \approx 1 - \langle b_k \rangle^{ME} + \frac{1}{2} \langle b_k^2 \rangle^{ME} \tag{8.39}$$

Here $\langle b_k^2 \rangle$ is evaluated directly from $\mathcal{P}_k(b_k)$ of Eq. (8.34) as

$$\langle b_k^2 \rangle = A_k L_k^{-3} [(x_1^2 + 2x_1 + 2)e^{-x_1} - (x_2^2 + 2x_2 + 2)e^{-x_2}] \tag{8.40}$$

8.7 SYNTHETIC DATA EXAMPLES

The Gibbs algorithm just described converges fairly rapidly for I(number of paths) $\lesssim$ K(number of unknown components)/2 and reasonable starting values for $\boldsymbol{\beta}^{(0)}$, and the $\langle q_k \rangle^{ME}$ are independent of these starting values. The algorithm was written in double precision to have at least 14 significant decimal digits accuracy on the VAX computer.

For test cases we assumed a grid of 10 components with nonfailure probabilities $q_1, \ldots, q_{10}$ on the four paths of nonfailure probabilities $Q_1, \ldots, Q_4$, as shown in Fig. 8.5. An x at the intersection of the vertical line for component k and the horizontal line for path i indicates that com-

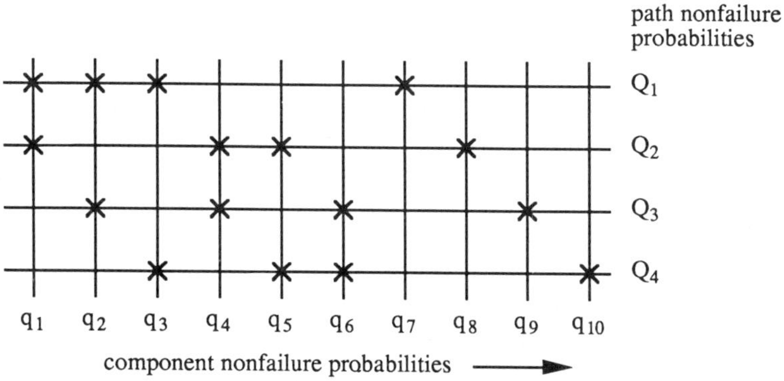

Figure 8.5 Connection diagram for ten components with nonfailure probabilities q_i (vertical lines) and five paths (horizontal lines). An x at the intersection of vertical line k and horizontal line i indicates C_k lies on path i: $m_{ik} = 1$. Otherwise $m_{ik} = 0$.

ponent C_k lies on path i: $m_{ik} = 1$. Otherwise $m_{ik} = 0$. Here, $I = 4$, $K = 10$.

Table 8.1 summarizes the results of a synthetic-data test case. All $\{q_k\}$ are bounded by (.99,1.). The associated $\{Q_i\}$ are .(2)98889, .(3)9699, .(2)98979, and .(2)97801. The $\langle q_k \rangle^{ME}$ remain unchanged if the bounds are changed to (.98,1.), except for a change of one unit in the last digit of $\langle q_5 \rangle^{ME}$ and $\langle q_{10} \rangle^{ME}$. The errors in these $\langle q_k \rangle^{ME}$ are, at most, one unit in the last digit.

What do we learn from a synthetic-data table like 8.1? Only that this choice of paths appears to estimate $q_{1,4,5,8}$ rather well and q_{10} rather poorly. If we had chosen a different set of four paths to estimate these $\{q_k\}$ of Fig. 8.5 we might have obtained better agreement with their $\langle q_k \rangle^{ME}$. In a large system with a practical limit to the number of test paths it is not usually obvious which choice of paths will yield the best ME estimates for a larger number of components on them.

TABLE 8.1 ME-Gibbs Estimates of Assumed $\{q_k\}$ for the Bounds (.99,1.)

Component k	1	2	3	4	5
q_k Assumed	.(4)900	.(5)90	.(3)9000	.(6)9	.(4)900
$\langle q_k \rangle^{ME}$	.(4)930	.(3)9781	.(3)9646	.(4)931	.(4)921
Component k	**6**	**7**	**8**	**9**	**10**
q_k Assumed	.(3)9000	.(6)9	.(4)900	.(5)90	.(4)900
$\langle q_k \rangle^{ME}$	.(3)9679	.(3)9532	.(4)917	.(3)9588	.(2)98555

The number in parentheses indicates the number of 9's; viz. .(3)9123 = .999123.

It appears that the primary advantage of this ME-Gibbs method of component assessment is that it offers a means of selecting via synthetic data examples, a group of $I \lesssim K/2$ paths which will estimate satisfactorily the nonfailure probabilities of K components in an ensemble of similar large systems. Then, after analyzing the path failure data any $\langle q_k \rangle^{ME}$ which is significantly lower than the anticipated value would indicate a possibly inferior component C_k.

If we didn't know q_k exactly, but could specify its bounds under actual operating conditions, we might accept the value $\langle q_k \rangle^{ME}$ obtained from path measurements as the *operating* nonfailure probability of C_k in an ensemble of similar systems.

REFERENCES

[8.1] C.L. Hwang, et al., System-reliability evaluation techniques for complex/large systems—A Review, IEEE Trans. Reliability, *R-30*, Dec. 1981, pp. 416–423.

[8.2] G.C. Corynen, A fast bottom-up algorithm for computing the cut sets of noncoherent fault trees, UCRL-53828, Lawrence Livermore National Laboratory, Nov., 1987.

Appendix A

Derivation of the Gibbs ME Distribution for a Spherical Earth Density, Eqs. (2.10) to (2.13)

We form the objective function F as the sum of various terms: h of Eq. (2.4), constraint Eq. (2.5) times Lagrange multiplier β_a, Eq. (2.7) multiplied by β_b, and Eq. (2.8) multiplied by β_c. F is

$$
\begin{aligned}
F = & - \int d\bar{\rho} p(\bar{\rho}) \log[p(\bar{\rho})] + \beta_a \left[\int d\bar{\rho} p(\bar{\rho}) - 1 \right] \\
& + \beta_b \left[\sum_n \int d\bar{\rho} \rho_n p(\bar{\rho}) - \rho_{av} N \right] \\
& + \beta_c \left[\sum_n \int d\bar{\rho} \rho_n [n^{5/3} - (n-1)^{5/3}] \, p(\bar{\rho}) - y \rho_{av} \, N^{5/3} \right]
\end{aligned}
\tag{A.1}
$$

where

$$
\int d\bar{\rho} = \int d\rho_1 \ldots \int d\rho_N, \qquad p(\bar{\rho}) = p(\rho_1, \ldots, \rho_N) \tag{A.2}
$$

We do not yet factor probability $p(\bar{\rho})$ into a product of N terms.

In general the mth constraint is

$$
\int d\bar{\rho} p(\bar{\rho}) F_m(\bar{\rho}) = \langle F_m \rangle \text{ observed} \tag{A.3}
$$

Setting $\delta F = 0$ with respect to an arbitrary continuous impulse-like variation $\delta p(\bar{\rho})$ at arbitrary $\bar{\rho}$, we get

$$0 = -\int d\bar{\rho}\{\delta p(\bar{\rho}) \log[p(\bar{\rho})] + \delta p(\bar{\rho})\} + \beta_a \int d\bar{\rho}\, \delta p(\bar{\rho}) + \beta_b \sum_n \int d\bar{\rho}\rho_n\, \delta p(\bar{\rho}) + \beta_c \sum_n \int d\bar{\rho}\rho_n[n^{5/3} - (n - 1)^{5/3}]\, \delta p(\bar{\rho}) \tag{A.4}$$

This rearranges to

$$0 = -\int d\bar{\rho}\, \delta p(\bar{\rho}) \left\{\log[p(\bar{\rho})] + 1 - \beta_a - \beta_b \sum \rho_n - \beta_c \sum \rho_n\, [n^{5/3} - (n - 1)^{5/3}]\right\} \tag{A.5}$$

Since this is true for an arbitrary functional change $\delta p(\bar{\rho})$ at arbitrary $\bar{\rho}$ the quantity within braces must be zero. Hence,

$$p(\bar{\rho}) = \exp\left\{- 1 + \beta_a + \sum_n (\beta_b + \beta_c\, [n^{5/3} - (n - 1)^{5/3}])\rho_n\right\} \tag{A.6}$$

The term $\exp[-1 + \beta_a]$ can be written as A. Rietsch [2.1] prefers to write $\beta_b = -\beta_1$, $\beta_c = -\beta_2/N^{2/3}$, so

$$p(\bar{\rho}) = p(\rho_1, \ldots, \rho_N) = A \exp\left[- \sum_n \rho_n\eta_n(\bar{\beta})\right] \tag{A.7}$$

with

$$\eta_n(\bar{\beta}) = \beta_1 + \beta_2 \frac{[n^{5/3} - (n - 1)^{5/3}]}{N^{2/3}} \tag{A.8}$$

To satisfy the normalization constraint, which makes the first [ ] in Eq. (A.1) zero, we evaluate A so

$$p(\bar{\rho}) = p(\rho_1, \ldots, \rho_N) = \frac{\exp\,[- \sum_n \rho_n\eta_n(\bar{\beta})]}{Z(\bar{\beta})} = \prod_n p_n(\rho_n) \tag{A.9}$$

where

$$Z(\bar{\beta}) = \int d\bar{\rho} \exp[- \sum_n \rho_n\eta_n(\bar{\beta})] = \prod_n Z_n(\bar{\beta}) \tag{A.10}$$

β_1 and β_2 are chosen to satisfy the constraints embodied in the second and third [ ] in Eq. (A.1) equal to zero.

Rietsch allows upper and lower bounds $\rho_l > 0$ and $\rho_u < N\rho_{av}$ for each

ρ_n, as follows. We evaluate $Z_n(\bar{\beta})$ in Eq. (A.10) as

$$Z_n(\bar{\beta}) = \int_{\rho_l}^{\rho_u} d\rho_n e^{-\rho_n \eta_n(\bar{\beta})} = \frac{e^{-\rho_l \eta_n \bar{\beta})}}{\eta_n(\bar{\beta})} (1 - e^{-(\rho_u - \rho_l)\eta_n(\bar{\beta})}) \qquad \text{(A.11)}$$

The constraint from the second bracketed expression in Eq. (A.1) is

$$\sum_n \langle \rho_n \rangle = \sum_n \int d\rho_n \rho_n p_n(\rho_n) = \rho_{av} N \qquad \text{(A.12)}$$

Substituting from Eq. (A.9) for $p_n(\rho_n) = \exp[-\rho_n \eta_n]/Z_n$ we get

$$\sum_n \left(\rho_l + \frac{1}{\eta_n(\bar{\beta})} - \frac{\rho_u - \rho_l}{\exp\,[(\rho_u - \rho_l)\eta_n(\bar{\beta})] - 1} \right) = \rho_{av} N \qquad \text{(A.13)}$$

This is one constraint on β_1 and β_2 in each $\eta_n(\bar{\beta})$ of Eq. (A.8). The other, from the third bracketed expression in Eq. (A.1), is

$$\sum_n \int d\rho_n \, \rho_n \, p_n \, (\rho_n) \, [n^{5/3} - (n - 1)^{5/3}] = y\rho_{av} N^{5/3} \qquad \text{(A.14)}$$

Here the summand is just the [ ] of Eq. (A.13) multiplied by $[n^{5/3} - (n - 1)^{5/3}]$.

Equations (A.13) and (A.14) may be readily solved by trial and error for β_1 and β_2, given the chosen bounds ρ_l and ρ_u, the selected number N of equivolume shells, and the data ρ_{av} and y of Eqs. (2.7) and (2.9). Then the entire probability function $p(\rho_1, \ldots, \rho_N) = \Pi\, p_n(\rho_n)$ is known in Eqs. (A.9) and (A.10). The ME estimates of the shell densities are the $\langle \rho_n \rangle$, evaluated as expectation values of the ρ_n over their probability distributions, by the integral in Eq. (A.12).

Appendix B

Derivation of the Gibbs ME Distribution for a Laterally Varying Earth Density

This derivation is an extension of the one in Appendix A for a spherically symmetric earth. If we set $\delta F = 0$ in Eq. (2.19) with respect to an arbitrary continuous impulse-like variation $\delta p(\bar{\rho})$ at arbitrary $\bar{\rho}$ we obtain

$$0 = -\int d\bar{\rho}\, \delta p(\bar{\rho})\{\log[p(\bar{\rho})] + 1\} + \beta_0 \int d\bar{\rho}\, \delta p(\bar{\rho}) + \sum_{q=1}^{Q} \beta_q \left[\sum_{n=1}^{N} \int d\bar{\rho}\rho_n\, \delta p(\bar{\rho}) f_q(\bar{r}_n)\right] \tag{B.1}$$

which rearranges into

$$0 = -\int d\rho\, \delta p(\bar{\rho})\Big\{\log[p(\bar{\rho})] + 1 - \beta_0 - \sum_q \beta_q \sum_n \rho_n f_q(\bar{r}_n)\Big\} \tag{B.2}$$

This being true for an arbitrary functional change $\delta p(\bar{\rho})$ at arbitrary $\bar{\rho}$, the quantity within braces must be zero. Hence,

$$p(\bar{\rho}) = \exp\left[-1 + \beta_0 + \sum_q \beta_q \sum_n f_q(\bar{r}_n)\rho_n\right] \tag{B.3}$$

Now define

$$\exp(-1 + \beta_0) = A, \qquad \sum_q \beta_q f_q(\bar{r}_n) = R_n(\bar{\beta}) \tag{B.4}$$

Then,

$$p(\overline{\rho}) = A \exp\left[-\sum_n R_n(\overline{\beta})\rho_n\right] \tag{B.5}$$

The constraint embodied in the first bracketed expression in Eq. (2.19) says

$$A\int d\overline{\rho}\exp\left[-\sum_n R_n(\overline{\beta})\rho_n\right] = 1, \qquad d\overline{\rho} = d\rho_1 \cdots d\rho_N \tag{B.6}$$

Assuming for simplicity no upper or lower bounds on each ρ_i we evaluate A by Eq. (B.6) as ΠR_n.

Therefore, it is logical to represent $p(\overline{\rho})$ in Eq. (B.5) as a product of factors for the N cells,

$$p(\overline{\rho}) = \prod_n p_n(\rho_n), \qquad p_n(\rho_n) = R_n e^{-R_n(\overline{\beta})\rho_n} \tag{B.7}$$

The qth constraint embodied in the second bracketed expression of Eq. (2.19) is, with this $p_n(\overline{\rho}_n)$,

$$\sum_{n=1}^{N}\int_0^\infty d\overline{\rho}\rho_n\left[\prod_m R_m e^{-R_m(\overline{\beta})\rho_m}\right] f_q(\overline{r}_n) = \langle F_q\rangle \tag{B.8}$$

For the nth term in the sum the integrals over $\rho_m \neq \rho_n$ are 1, so

$$\sum_n \frac{f_q(\overline{r}_n)}{R_n(\overline{\beta})} = \langle F_q\rangle, \qquad q = 1, \ldots, Q \tag{B.9}$$

This set of Q equations involving the R_n $(\beta_1, \ldots, \beta_Q)$ determines implicitly the Q values of the β_i.

Once all these $\beta_1, \ldots, \beta_Q$ are found from the measured gravity field coefficients F_q all the $R_n(\overline{\beta})$ are known from Eq. (B.4). Then, the entire probability function $p(\rho_1, \ldots, \rho_n) = \Pi p_n(\rho_n)$ is known in Eq. (B.7). The ME estimates of the cell masses are the $\langle\rho_n\rangle$, computed as

$$\langle\rho_n\rangle = \int d\overline{\rho}\rho_n p(\overline{\rho}) = R_n\int d\rho_n\rho_n e^{-R_n(\overline{\beta})\rho_n} = \frac{1}{R_n(\overline{\beta})} \tag{B.10}$$

This $\langle\rho_n\rangle = \langle\rho_n^{\text{ME}}\rangle$ is *not* Eq. 13 of reference [2.3].

Appendix C

Statistical Distribution of a Real-Valued Parameter by the Darwin-Fowler Method [3.7]

We obtain the statistical distribution of an object parameter x_k about its average value, which is its ME value x_k^{ME}, in an ensemble of possible objects. Each such object is described by the vector $\overline{x} = (x_1, \ldots, x_k)$ and $\sum x_k = x_{\text{T}}$ is constrained. The measurement constraints are embodied in the matrix equation $\mathbf{Dx} = \mathbf{T}$ of Eq. (3.3).

In order to use the D-F method [C.1] we must quantize $\overline{x}$. This we do by introducing a resolution level $\hat{d}x$, dependent on the measurement system and the characteristics of the true object but independent of any sources of noise. We also introduce a reference D_r and reference $\hat{d}\text{T}$ determined by D_r in the scalar equation

$$\text{D}_r \hat{d}x = \hat{d}\text{T} \tag{C.1}$$

Now divide $\mathbf{Dx} = \mathbf{T}$ by Eq. (C.1),

$$\mathbf{D'n} = \mathbf{T'}, \qquad \mathbf{n} = \left[\frac{\mathbf{x}}{\hat{d}x}\right] \tag{C.2}$$

Integer matrix $\mathbf{n}$ now describes the object; the brackets denote the integer value n_k of each $x_k/\hat{d}x$. $\mathbf{D'} = \mathbf{D}/\mathbf{D}_r$ and $\mathbf{T'} = \mathbf{T}/\hat{d}\text{T}$ are not integerized.

Each of the I constraint relations cannot be satisfied *exactly* and so we relax the ith one as

$$-\Delta_i \le \left(\sum_k \text{D}'_{ik} n_k - \text{T}'_i\right) \le +\Delta_i, \qquad \text{all } i \le \text{I} \tag{C.3}$$

If D_r is chosen sufficiently large and $\hat{d}T$ along with it, all the D'_{ik} and T'_i will be sufficiently small and Eq. (C.3) will be satisfied for small Δ_i. Fortunately the Δ_i parameters do not affect the final statistical distribution of each n_k!

We now derive an integral expression which will yield the moments of a given n_α relative to its ensemble-average value $\langle n_\alpha \rangle$. The total number of microstates which can realize a macrostate $(n_1, \ldots, n_k)$ is [see Eq. (1.3) for the Maxwell-Boltzmann distribution]

$$C = \sum_{n_\alpha}{}' \sum_{\substack{\text{all} \\ n_k \neq n_\alpha}}{}' \frac{N!}{n_\alpha!\left(\prod_{k\neq\alpha} n_k!\right)} \tag{C.4}$$

The primes indicate the constraints of $\sum n_k = N$ (fixed) and Eq. (C.3). In what follows, we dispense with the primes in D'_{ik} and T'_i in Eq. (C.3) for convenience. The desired integral expression depends on the identity

$$(z_1 + z_2 + \cdots + z_K)^N = \sum_{\text{sets}}{}' \frac{N!}{\prod_k n_k!} z_1^{n_1} \cdots z_K^{n_K} \tag{C.5}$$

The sets consist of all arrangements of the $\{n_k\}$; we want to count only those which satisfy constraints Eq. (C.3).

Now, for reasons to become evident, we change

$$\begin{array}{lll} z_1 & \text{to} & \exp(j\,\rho_1 D_{11}) \exp(j\rho_2 D_{21}) \cdots \exp(j\rho_I D_{I1}) \\ \vdots & & \vdots \\ z_k & \text{to} & \exp[j \sum_i \rho_i D_{ik}] \qquad j = \sqrt{-1} \\ \vdots & & \vdots \end{array} \tag{C.6}$$

Then Eq. (C.5) can be written

$$\begin{aligned} Z^N(\{\rho_k\}) &= \left[\exp\left(j\sum_{i=1}^{I} \rho_i D_{i1}\right) + \cdots + \exp\left(j\sum_i \rho_i D_{ik}\right) + \cdots\right]^N \\ &= \sum_{\text{sets}} \frac{N!}{\prod n_k!} \exp\left(j\rho_1 \sum_k D_{1k} n_k\right) \cdots \exp\left(j\rho_i \sum_k D_{ik} n_k\right) \cdots \end{aligned} \tag{C.7}$$

In order to remove all the factors on the right and constrain the sum we utilize the Dirichlet factor

$$f_i = \frac{1}{\pi} \int_{-\infty}^{\infty} \frac{\sin(\Delta_i \rho_i)}{\rho_i} \exp\left[j\rho_i\left(\sum_k D_{ik} n_k - T_i\right)\right] d\rho_i \tag{C.8}$$

which has the value 1 when the ( )-term in Eq. (C.3) is in the $\pm\Delta_i$-range and zero otherwise. Thus, f_i is 1 only when the ith constraint in Eq. (C.4) is satisfied.

To place f_i-factors on the right side of Eq. (C.7) multiply both sides by

$$\prod_i \frac{\sin(\Delta_i \rho_i)}{\rho_i} e^{-j\rho_i T_i} \tag{C.9}$$

to obtain

$$Z^N(\{\rho_i\}) \left(\prod_i \frac{\sin(\Delta_i \rho_i)}{\rho_i} e^{-j\rho_i T_i}\right)$$

$$= \sum_{\text{sets}} \frac{N!}{\prod n_k!} \left(\prod_i \frac{\sin(\Delta_i \rho_i)}{\rho_i} \exp\left[j\rho_i\left(\sum_k D_{ik} n_k - T_i\right)\right]\right) \tag{C.10}$$

Now, on taking $\Pi_i\left(\frac{1}{\pi}\int d\rho_i\right)$ of Eq. (C.10) we obtain the right side constrained with the f_j

$$\left(\frac{1}{\pi}\right)^I \int_{-\infty}^{\infty} d\rho_1 \frac{\sin(\Delta_1 \rho_1)}{\rho_1} e^{-j\rho_1 T_1} \cdots \int_{-\infty}^{\infty} d\rho_I \frac{\sin(\Delta_I \rho_I)}{\rho_I} e^{-j\rho_I T_I} Z^N(\{\rho_i\})$$

$$= \sum_{\text{sets}}{}' \frac{N!}{\prod n_k!} \equiv C \text{ of Eq. (C.4)} \tag{C.11}$$

The essence of the D-F method is that the left side of Eq. (C.11) has properties which expedite the computation of moments of any n_α about its $\langle n_\alpha \rangle$ *without explicit evaluation of the integrals*!

Note that $Z^N(\{\rho_i\})$ in Eq. (C.7) contains many terms of rapidly varying phase and their sum is raised to an enormous power N. This implies much cancellation of the integrands in Eq. (C.11) over the ρ_i paths along the real axis. If a typical ρ_i path is deformed into the complex plane near the origin, either upward along a semicircle to a maximum of $\rho_i = +j|\beta_i|$ or downward along a semicircle to a minimum of $-j|\beta_i|$, most of the contribution in the multiple integral Eq. (C.11) will come from the $+j\beta_i$ regions. We chose that set of deformed paths $\{P_i\}$ which maximizes Eq. (C.11).

We use these properties of Eq. (C.11) in evaluating $\langle n_\alpha \rangle$ by Eq. (C.4),

$$\langle n_\alpha \rangle = \frac{\displaystyle\sum_{n_\alpha}{}' n_\alpha \sum_{n_k \neq n_\alpha}{}' \frac{N!}{\left(\prod_{k\neq\alpha} n_k!\right) n_\alpha!}}{C} \tag{C.12}$$

The numerator is

$$N \sum_{n_\alpha}{}' \sum_{n_k \neq n_\alpha}{}' \frac{(N-1)!}{\left(\prod_{k\neq\alpha} n_k!\right)(n_\alpha - 1)!} \tag{C.13}$$

The double sum is of the same form as C of Eq. (C.4) except that (N − 1) replaces N and $(n_\alpha - 1)$ replaces n_α.

Since the constraints $\sum n_k = \mathrm{N}$ and Eq. (C.3) can be written

$$\sum_{k\neq\alpha} n_k + (n_\alpha - 1) = \mathrm{N} - 1 \tag{C.14}$$

$$-\Delta_i \le \sum_{k\neq\alpha} \mathrm{D}_{ik} n_k + \mathrm{D}_{i\alpha}(n_\alpha - 1) - (\mathrm{T}_i - \mathrm{D}_{i\alpha}) \le +\Delta_i, \quad i \le \mathrm{I} \tag{C.15}$$

we can rewrite Eq. (C.13) by inspection, by analogy with Eq. (C.11) for C, as

$$\begin{aligned}\text{numerator } \langle n_\alpha\rangle = \mathrm{N} &\int_{-\infty}^{\infty} d\rho_1 \frac{\sin(\Delta_1\rho_1)}{\rho_1} e^{-j\rho_1(\mathrm{T}_1 - \mathrm{D}_{1\alpha})} \\ \cdots &\int_{-\infty}^{\infty} d\rho_\mathrm{I} \frac{\sin(\Delta_\mathrm{I}\rho_\mathrm{I})}{\rho_\mathrm{I}} e^{-j\rho_\mathrm{I}(\mathrm{T}_1 - \mathrm{D}_{\mathrm{I}\alpha})} \mathrm{Z}^{\mathrm{N}-1}(\{\rho_i\})\end{aligned} \tag{C.16}$$

The only difference between the integrands of Eqs. (C.16) and (C.11) is the factor

$$\frac{e^{j\rho_1\mathrm{D}_{1\alpha}} \cdots e^{j\rho_\mathrm{I}\mathrm{D}_{\mathrm{I}\alpha}}}{\mathrm{Z}(\{\rho_i\})} \tag{C.17}$$

Since this is slowly varying in the important region $\{\rho_i\} = \{j\beta_i\}$ we can remove it. The remarkably simple result is that Eq. (C.16) becomes

$$\frac{num\ \langle n_\alpha\rangle}{\mathrm{C}} = \langle n_\alpha\rangle = \mathrm{N}\left[\frac{\exp\left(\sum_i \beta_i \mathrm{D}_{i\alpha}\right)}{\mathrm{Z}(\{\beta_i\})}\right] \tag{C.18}$$

This is the maximum-entropy form Eq. (3.10) for $n_\alpha^{\mathrm{ME}} = x_\alpha^{\mathrm{ME}}/\hat{dx}$. This result supports the reasoning that if C of Eq. (C.11) and *num* $\langle n_\alpha\rangle$ of Eq. (C.16) are evaluated along "stationary phase" paths in the $\{\rho_i\}$ planes, the result will be the most probable, maximum entropy result, consistent with the constraints.

We shall assume that the I-multiple integral in Eq. (C.16) is carried out along paths $\{\mathrm{P}_i\}$ so chosen to yield $\langle n_\alpha\rangle = n_\alpha^{\mathrm{ME}}$ for a maximum-entropy $\{n_i\}$ solution. The D-F method allows us to obtain higher order moments of n_α by this procedure for $\langle n_\alpha\rangle$ *independent* of the details of the I multiple integration! The result will be a Poisson-like distribution for n_α, which, as N → ∞, becomes a pure Poisson distribution with parameter $\langle n_\alpha\rangle$.

The next moment of n_α is

$$\langle n_\alpha^2\rangle = \frac{\displaystyle\sum_{n_\alpha}{}' n_\alpha^2 \sum_{n_k\neq n_\alpha}{}' \frac{\mathrm{N}!}{\left(\prod_{k\neq\alpha} n_k!\right) n_\alpha!}}{\mathrm{C}} \tag{C.19}$$

The numerator reduces to

$$\text{numerator } \langle n_\alpha^2 \rangle = N(N-1) \sum_{n_\alpha}{}' \sum_{n_k \neq n_\alpha}{}' \frac{(N-2)!}{\left(\prod_{k\neq\alpha} n_k!\right)(n_\alpha - 2)!} + \langle n_\alpha \rangle C \tag{C.20}$$

If we rewrite the constraints as

$$\sum_{k\neq\alpha} n_k + (n_\alpha - 2) = N - 2 \tag{C.21}$$

$$-\Delta_i \leq \left[\sum_{k\neq\alpha} D_{ik} n_k + D_{i\alpha}(n_\alpha - 2) - (T_i - 2D_{i\alpha}) \right] \leq +\Delta_i, \quad i \leq I \tag{C.22}$$

we can evaluate the double sums in Eq. (C.20) by analogy with C of Eq. (C.11) by replacing N by (N − 2) and all T_i by $(T_i - 2D_{i\alpha})$. The new multiple integral is the same as for C except for the additional factor

$$\frac{e^{j2\rho_1 D_{1\alpha}} \ldots e^{j2\rho_I D_{I\alpha}}}{Z^2(\{\rho_i\})} \tag{C.23}$$

Upon removing this factor, which is slowly varying in the region $\{\rho_i = j\beta_i\}$ we obtain for Eq. (C.20)

$$\text{numerator } \langle n_\alpha^2 \rangle = N(N-1) \left(\frac{e^{j\rho_1 D_{1\alpha}} \cdots e^{j\rho_I D_{I\alpha}}}{Z(\{\rho_i\})} \right)^2 C + \langle n_\alpha \rangle C \tag{C.24}$$

This, using Eq. (C.18), yields for the variance

$$\langle (n_\alpha - \langle n_\alpha \rangle)^2 \rangle = \langle n_\alpha^2 \rangle - \langle n_\alpha \rangle^2 = \langle n_\alpha \rangle^2 \left(\frac{1}{\langle n_\alpha \rangle} - \frac{1}{N} \right) \tag{C.25}$$

Similarly, the third moment is

$$\langle (n_\alpha - \langle n_\alpha \rangle)^3 \rangle = \langle n_\alpha^3 \rangle \left(\frac{1}{\langle n_\alpha \rangle} - \frac{1}{N} \right) \left(\frac{1}{\langle n_\alpha \rangle} - \frac{2}{N} \right) \tag{C.26}$$

Generalizing, it is easy to prove

$$\langle n_\alpha (n_\alpha - 1) \cdots (n_\alpha - m) \rangle = N(N-1) \cdots (N-m) \left(\frac{\langle n_\alpha \rangle}{N} \right)^{m+1} \tag{C.27}$$

In the limit $N \to \infty$ the right side becomes $\langle n_\alpha \rangle^{m+1}$, which is precisely the result for the Poisson distribution [C.2]

$$\lim_{N\to\infty} \text{Prob}\,(n_\alpha) = \frac{\langle n_\alpha \rangle^{n_\alpha}}{n_\alpha!} e^{-\langle n_\alpha \rangle}, \qquad \langle n_\alpha \rangle = n_\alpha^{\text{ME}} \tag{C.28}$$

This is the statistical distribution of n_α among all the microstates C of Eq. (C.4) which satisfy the constraints $\sum n_k = N$ and Eq. (C.3). The average value of any n_α is just n_α^{ME} in the distribution Eqs. (C.27) or (C.28).

REFERENCES

[C.1] R.B. Lindsay and H. Margenau, *Foundations of Physics,* New York: Dover, 1957, pp. 252–279.

[C.2] A.M. Mood, *Introduction to the Theory of Statistics,* New York: McGraw Hill, 1950, p. 59.

Appendix D

Derivation of the Maximum Entropy Complex Multivariate Gaussian Probability Density

The variable space consists of complex x_α at discrete points $\alpha = 1, \ldots, N + 1$. The probability of the set $\{x_\alpha, x_\alpha^*\}$ is denoted $P(x_1, x_1^*; \ldots, x_{N+1}, x_{N+1}^*)$, the asterisk (*) denoting complex conjugate. For brevity let $\mathbf{X}$ denote the space of $\{x_\alpha, x_\alpha^*\}$. We assume each x_α has no average: $\int \cdots \int x_\alpha P(\mathbf{X})D(\mathbf{X}) = 0$, where $D(\mathbf{X}) = \Pi\, dx_i dx_i^*$. We will maximize (real) entropy h defined over $\mathbf{X}$ as

$$h = - \int \cdots \int P(\mathbf{X}) \log[P(\mathbf{X})]\, D(\mathbf{X}) \tag{D.1}$$

subject to the constraints

$$\int \cdots \int P(\mathbf{X})D(\mathbf{X}) = 1 \tag{D.2}$$

and the measured complex autocorrelations R_i for $0 \le i \le p$, $p \le N$,

$$\langle x_\alpha^* x_{\alpha+i} \rangle = R_i = \int \cdots \int P(\mathbf{X}) \left[\frac{1}{p + 1 - i} \sum_{\alpha=1}^{N+1-i} x_\alpha^* x_{\alpha+i} \right] D(\mathbf{X}) \tag{D.3}$$

This is not an ergodic statement about the *process*; rather, an imposition of equivalence between each space-average autocorrelation of the process on the left and its ensemble-average representation on the right. (Factor $(p + 1 - i)^{-1}$ appears in both space- and ensemble-average.)

The objective function which we extremize by varying all the x_α and x_α^* (i.e., real and imaginary parts) independently is

$$
\begin{aligned}
F = h + \gamma\left[\int \cdots \int P(\mathbf{X})D(\mathbf{X}) - 1\right] \\
- \sum_{i=0}^{p} \beta_i' \left\{ \int \cdots \int P(\mathbf{X}) \left[\frac{1}{p+1-i} \sum_{\alpha} x_\alpha^* x_{\alpha+i} \right] D(\mathbf{X}) - R_i \right\} \\
- \sum_{i=1}^{p} \beta_i'^* \left\{ \int \cdots \int P(\mathbf{X}) \left[\frac{1}{p+1-i} \sum_{\alpha} x_\alpha x_{\alpha+i}^* \right] D(\mathbf{X}) - R_i^* \right\}
\end{aligned}
\tag{D.4}
$$

in terms of a set of complex Lagrange multipliers $\{\beta_i', \beta_i'^*\}$, for $0 \leq i \leq$ p.

If we define $\beta_i = \beta_i'/(p + 1 - i)$, $\beta_i^* = \beta_i'^*/(p + 1 - i)$ and set $\delta F = 0$ for a change δP of P within a small range of x_α^* for arbitrary α around some value we obtain from Eq. (D.4)

$$
\begin{aligned}
0 = \int \cdots \int \Bigg[-\log P - 1 + \gamma - \sum_{i=0}^{p} \sum_{\alpha} \beta_i x_\alpha^* x_{\alpha+i} \\
- \sum_{i=1}^{p} \sum_{\alpha} \beta_i^* x_\alpha x_{\alpha+i}^* \Bigg] \delta P(\mathbf{X}) D(\mathbf{X})
\end{aligned}
\tag{D.5}
$$

But the $\sum\sum$-terms integrate out over all $x_{\alpha+i}$ and x_α because we assumed zero averages for them. Eq. (D.5) tells us only that log P equals $-1 + \gamma +$ terms which integrate out, and therefore, are not linear in any of the x_α^*.

If we set $\delta F = 0$ for arbitrary δP in a narrow range of coordinates of the quadratic $x_\alpha^* x_{\alpha+j}$ for arbitrary α and j Eq. (D.5) tells us that only the $x_\alpha^* x_{\alpha+j}$ term in the sums will not be integrated out, and therefore,

$$
\begin{aligned}
\log P = {} & -1 + \gamma - \beta_j x_\alpha^* x_{\alpha+j} \\
& + \text{linear terms excluding } x_\alpha^* \text{ and } x_{\alpha+j} \\
& + \text{quadratic terms excluding } x_\alpha^* x_{\alpha+j} \\
& + \text{triplet, quadruplet, } \cdots \text{ products}
\end{aligned}
\tag{D.6}
$$

Since Eq. (D.6) must be true for *all combinations* of x_α^* and $x_{\alpha+j}$, and also all x_α and $x_{\alpha+j}^*$, we can conclude there are no linear terms and

$$
\begin{aligned}
\log P = -1 + \gamma - \sum_{i=0} \sum_{\alpha} \beta_i x_\alpha^* x_{\alpha+i} - \sum_{i=1} \sum_{\alpha} \beta_i^* x_\alpha x_{\alpha+i}^* \\
+ \text{possible higher} - \text{order products}
\end{aligned}
\tag{D.7}
$$

If we take a variation such that $\delta P \neq 0$ at the coordinates of any higher order product, δF would be nonzero. Therefore, all higher order products are absent and we write

$$
\begin{aligned}
P(\mathbf{X}) &= B_0 \exp\left[-\sum_{i=0}^{p} \beta_i \sum_{\alpha=1}^{N+1-i} x_\alpha^* x_{\alpha+i} - \sum_{i=1}^{p} \beta_i^* \sum_{\alpha} x_\alpha x_{\alpha+i}^*\right] \\
&= B_0 \exp(-\mathbf{x}^\dagger \boldsymbol{\beta} \mathbf{x}) \qquad \text{(D.8)}
\end{aligned}
$$

Here, $B_0 = e^{-1+\gamma}$, $\mathbf{x}$ = column $(x_1, \ldots, x_N)$, $\mathbf{x}^\dagger$ = row $(x_1^*, \ldots, x_N^*)$, and $\boldsymbol{\beta}$ is the $(N + 1) \times (N + 1)$ Hermitian matrix

$$
\boldsymbol{\beta} = \begin{array}{c} \\ 1 \\ 2 \\ \\ \\ \\ \\ p+1 \\ \\ \\ \\ \\ N+1 \end{array}
\begin{array}{c}
\begin{array}{ccccccccccccc} 1 & 2 & & & & & p+1 & & & & & & N+1 \end{array} \\
\left(\begin{array}{ccccccc|cccccc}
\beta_0 & \beta_1 & \cdots & \cdots & \cdots & \cdots & \beta_p & 0 & \cdots & \cdots & & 0 & 0 \\
\beta_1^* & \beta_0 & \beta_1 & \cdots & \cdots & \cdots & \beta_{p-1} & \beta_p & 0 & \cdots & & \cdots & 0 \\
\vdots & \vdots & & & & & \vdots & \vdots & & & & \vdots & \vdots \\
\beta_{N-p-1}^* & \beta_{N-p-2}^* & \cdots & \cdots & \cdots & \cdots & \vdots & \vdots & \cdots & \beta_{p-1} & & \beta_p & 0 \\
\beta_{N-p}^* & \beta_{N-p-1}^* & & & \cdots & \cdots & \vdots & \vdots & \cdots & \cdots & & \beta_{p-1} & \beta_p \\
\vdots & \vdots & & & & & \vdots & \vdots & & & & & \vdots \\
\beta_p^* & \beta_{p-1}^* & \cdots & \cdots & \cdots & \cdots & \beta_0 & \beta_1 & \cdots & \cdots & & \cdots & \beta_{N-p} \\
\hline
0 & \beta_p^* & \beta_{p-1}^* & \cdots & \cdots & \cdots & \cdots & \beta_0 & \cdots & \cdots & & \cdots & \beta_{N-p-1} \\
\vdots & & & & & & & \vdots & & & & & \vdots \\
\vdots & \ddots & & & & & & \vdots & & & & & \vdots \\
0 & \cdots & \ddots & \beta_p^* & \beta_{p-1}^* & \cdots & \cdots & \cdots & \cdots & \beta_{-1}^* & & \beta_0 & \beta_1 \\
0 & 0 & \cdots & 0 & \beta_p^* & \cdots & \cdots & \cdots & \cdots & \cdots & & \beta_1^* & \beta_0
\end{array}\right)
\end{array}
\qquad \text{(D.9)}
$$

If we define the $(N + 1) \times (N + 1)$ Hermitian autocorrelation matrix $\mathbf{R}_N$ as

$$\mathbf{R}_N = \boldsymbol{\beta}^{-1} \qquad \text{(D.10)}$$

and choose the $p + 1$ elements of $\boldsymbol{\beta}$ to yield the measured $R_0, \ldots, R_p$ according to Eq. (D.3) then Eq. (D.8) yields the complex multivariate Gaussian

$$P(\mathbf{X}) = B_0 \exp(-\mathbf{x}^\dagger \mathbf{R}_N^{-1} \mathbf{x}) \qquad \text{(D.11)}$$

which is normalized according to Eq. (D.2) for

$$B_0 = (\pi^{N+1} \, \| \mathbf{R}_N \|)^{-1} \qquad \text{(D.12)}$$

$\| \mathbf{R}_N \|$ being the (real) determinant of $\mathbf{R}_N$. The elements $R_{p+1}, \ldots, R_N$ are the ME-estimated values.

If all the x_α are real the matrix Eq. (D.10) is real and the normalized density $P(x_1, \ldots, x_N)$ is proportional to the square root of Eq. (D.11)

$$P(x_1, \ldots, x_N) = [(2\pi)^{(N+1)/2} \, \| \mathbf{R}_N \|^{1/2}]^{-1} \exp\left[-\frac{1}{2} \mathbf{x}^t \mathbf{R}_N^{-1} \mathbf{x}\right] \qquad \text{(D.13)}$$

$\mathbf{x}^t$ is row $(x_1, \ldots, x_N)$.

Appendix E

The Entropy of Failure Probabilities of a Number of Independent System Components

We have K components $C_1, \ldots, C_K$ and the ith one has failure probability p_i independent of all the others. We will derive the entropy of all the failure and nonfailure states of all the components.

First assume just two components, C_1 and C_2, and an ensemble of N failure and nonfailure states among them. There are four overall-states: fail1-fail2, fail1-nonfail2, nonfail1-fail2, and nonfail1-nonfail2. In a macrostate of all these specified by p_1 and p_2 there are, respectively, $p_1p_2\mathrm{N}$, $p_1(1 - p_2)\mathrm{N}$, $(1 - p_1)p_2\mathrm{N}$, and $(1 - p_1)(1 - p_2)\mathrm{N}$ components. The number of combinations (microstates) C for this macrostate is

$$\mathrm{C} = \frac{\mathrm{N}!}{(p_1p_2\mathrm{N})! \cdots [(1 - p_1)(1 - p_2)\mathrm{N}]!} \tag{E.1}$$

This is the number of independent ways of distributing N objects into four bins, holding $(p_1 p_2, \mathrm{N}), \ldots, (1 - p_1)(1 - p_2)\mathrm{N}$ objects. The entropy log(C) is, with reference to Eqs. (1.1) to (1.6),

$$\begin{aligned} \mathrm{H} = \log \mathrm{C} \approx -\,\mathrm{N}\{&p_1p_2 \log(p_1p_2) + p_1(1 - p_2)\log[p_1(1 - p_2)] \\ &+ (1 - p_1)p_2 \log[(1 - p_1)p_2] \\ &+ (1 - p_1)(1 - p_2)\log[(1 - p_1)(1 - p_2)]\} \end{aligned} \tag{E.2}$$

Collecting the $\log p_1$, $\log(1 - p_1)$, $\log p_2$, and $\log(1 - p_2)$ terms, we obtain

$$\begin{aligned} \mathrm{H} = -\mathrm{N}[&p_1 \log p_1 + (1 - p_1)\log(1 - p_1) \\ &+ p_2 \log p_2 + (1 - p_2)\log(1 - p_2)] \end{aligned} \tag{E.3}$$

Generalizing this derivation to more than two independent components yields

$$H = -N \sum_k [p_k \log p_k + (1 - p_k) \log(1 - p_k)] \tag{E.4}$$

The entropy h per state is

$$h = -\sum_k [p_k \log p_k + (1 - p_k) \log(1 - p_k)] \tag{E.5}$$

which is Eq. (8.15) of the text, summed over all the components.

Index

A

B

C

D

N

S

T

U

V

W

Y

Z